**The Essential Guide to
Analytical Chemistry**

The Essential Guide to Analytical Chemistry

Translation of the revised and updated German Second Edition

Georg Schwedt

Institute of Inorganic and Analytical Chemistry, Technical University Clausthal, Clausthal-Zellerfeld, Germany

Translated by **Brooks Haderlie**

JOHN WILEY & SONS
Chichester · New York · Weinheim · Brisbane · Singapore · Toronto

iv

Title of the original edition in German: *Taschenatlas der Analytik*, 2nd Edition, 1996, © Georg Thieme Verlag, Stuttgart–New York

English language translation copyright © 1997 John Wiley & Sons Ltd,
Baffins Lane, Chichester,
West Sussex PO19 1UD, England

National 01243 779777
International (+ 44) 1243 779777
e-mail (for orders and customer service enquiries): cs-books@wiley.co.uk

Visit our Home Page on http://www.wiley.co.uk or http://www.wiley.com

Other Wiley Editorial Offices

John Wiley & Sons, Inc., 605 Third Avenue,
New York, NY 10158-0012, USA

WILEY-VCH Verlag GmbH, Pappelallee 3,
D-69469 Weinheim, Germany

Brisbane · Singapore · Toronto

Library of Congress Cataloging-in-Publication Data
Schwedt, Georg, 1943–
[Taschenatlas der Analytik. English]
The essential guide to analytical chemistry / Georg Schwedt;
translated by Brooks Haderlie.
p. cm.
'A translation of a revised and updated second edition.'
Includes bibliographical references (p. –) and index.
ISBN 0-471-97412-9 (pbk.)
1. Chemistry, Analytic. I. Title.
QD75.2.S3913 1997
543—dc21 97-2833
CIP

British Library Cataloguing in Publication Data
A catalogue record for this book is available from the British Library

ISBN 0 471 97412 9

Typeset in Times Roman by Techset Composition, Salisbury, UK.
Printed and bound in Great Britain by Bath Press Colourbooks, Glasgow

Contents

Preface to the First German Edition vii

Preface to the Second German Edition ix

Preface to the English Edition xi

Introduction 1

1 Basic Principles 2
  1.1 Goals of Chemical Analysis 2
  1.2 Formulation of the Problem and Analytical Strategy 4
  1.3 Classification of Analytical Methods 6
  1.4 Fundamental Working Steps and Methods in Synthesis 8
  1.5 Fields of Application and Comparison of Analytical Methods 10
  1.6 Composite Methods and Sources of Errors 12
  1.7 Statistical Evaluation of Analytical Results 14

2 Sample Preparation 18
  2.1 Sampling and Sample Stabilization 18
  2.2 Decomposition Methods 22
  2.3 Enrichment Methods 28
  2.4 Sample Cleanup Procedures 30

3 Detection Methods 32
  3.1 Working Methods and Analytical Process 32
  3.2 Separation Procedures and Selective Reagents 34
  3.3 Special Test Methods 36

4 Chemical and Biochemical Methods 38
  4.1 Gravimetry 38
  4.2 Titrimetric Analysis 40
  4.3 Enzymic Analysis 44
  4.4 Immunochemical Methods 48

5 Electrochemical Analytical Methods 50
  5.1 Basic Information 50
  5.2 Electrogravimetry 52
  5.3 Potentiometry 54
  5.4 Conductometry 60
  5.5 Polarography and Voltammetry 62
  5.6 Coulometry 70

6 Thermal Analytical Methods 72
  6.1 Methods—Overview 72
  6.2 Thermogravimetry 74
  6.3 Differential Thermal Analysis 76

6.4 Differential Scanning Calorimetry   80

7 Atomic Spectrometric Methods   82
7.1 Atomic Absorption Spectrometry   82
7.2 Atomic Emission Spectrometry   92
7.3 X-ray Fluorescence Analysis   98

8 Molecular Spectroscopic Analytical Methods   104
8.1 Introduction   104
8.2 Colorimetry   106
8.3 Spectrophotometry   108
8.4 Fluorimetry   112
8.5 Infrared and Raman Spectroscopy   114
8.6 Mass Spectrometry   122
8.7 Nuclear Magnetic Resonance Spectroscopy   132

9 Separation Methods   140
9.1 Systematic Physical and Chemical Separation Methods   140
9.2 Chromatographic Separation Methods   142
9.3 Electrophoresis   174

10 Automation of Analytical Procedures   183
10.1 Continuous Flow Analysis   186
10.2 Flow Injection Analysis   190
10.3 Coupling Techniques   196

11 Special Fields of Application and Methods   200
11.1 Radiochemical Methods   200
11.2 Solid Body and Surface Analysis   202
11.3 Chemical Sensors   206
11.4 Process Analysis   210
11.5 Structural Analysis   216
11.6 Elemental Species Analysis   220
11.7 Water Analysis   222
11.8 Analysis of Aroma Substances   224
11.9 Pesticide or Residue Analysis   226

Bibliography   230

Index   232

Preface to the First German Edition

The intended audience for this handbook includes all scientists who are involved with materials analysis—from participants in chemistry courses at the secondary level of continuing education schools, to students at technical and scientific colleges, and on to the specialists, those technically competent users of analytical chemistry data whose task at hand is problem-oriented interpretation. They are often the executors of analyses and, in consultation with the analyst, they should already have an idea of the possibilities and limits of the methods. Only those who view chemical analysis as a whole—from the formulation of the problem and sample preparation, to the application of methods and on to a critical interpretation of the experimental values—will meet the challenge of providing meaningful, usable data in view of the high expense which results from the use of first-rate analytical systems and qualified personnel.

During the conception of this work, and in very close, constructive cooperation with the graphic artist Joachim Schreiber, I established the goal of presenting in words and pictures the most important methods of current chemical-physical analysis, from their basic principles to problem-oriented applications. This meant dealing with a cross-section which would be as broad as possible. Since the pictorial representation was of prime importance, I often had to compromise with the text, which had to be kept concise. Therefore, this handbook cannot and should not replace a textbook. Numerous textbooks and special monographs from many colleagues served as a pattern for the graphical representations. These references are cited in the Appendix.

I hope that this volume will be well received by experts in the field, and I would appreciate comments and additional suggestions.

Clausthal-Zellerfeld
Spring 1992
Georg Schwedt

Preface to the Second German Edition

The Essential Guide to Analytical Chemistry (German title: *Taschenatlas der Analytik*), which was first published in the summer of 1992 and which has been translated into French, Japanese and Greek, was fortunately so well received that a new edition had to be prepared after only three years. Typographical and technical errors were corrected in the second edition, and three new color plates were added: The 'Structural Analysis' chapter has a new section on diffraction methods, and a chapter on 'Chemical Sensors' has been added (biochemical sensors were treated briefly in the first edition in Section 5.3 under Electrodes). The figures for these three plates come from the author's textbook *Analytische Chemie. Grundlagen, Methoden und Praxis* (Stuttgart, 1995), which contains all color figures and which was also published by Georg Thieme.

Clausthal-Zellerfeld
Fall/Winter 1995/96   Georg Schwedt

Preface to the English Edition

After being published in French, Japanese and Greek, this *Essential Guide to Analytical Chemistry* is now also available in English. I hope this edition will be as well received as the second German edition was. Compared with the first edition, it has been supplemented with a new chapter on chemical sensors, and the chapter on structural analysis has a new section on diffraction methods. I want to thank Mr Brooks Haderlie for translating the manuscript from German into English and all at John Wiley & Sons for their work.

Spring 1996   George Schwedt

Important Notice

This volume has been compiled by specialists. The user must know that simply working with chemicals and microorganisms involves latent risks. Theoretically, additional risks can result from improper amounts of materials. The authors, editors and publishers took great care to ensure that the amounts and experimental protocols met the current state of scientific art when this volume was published. However, in spite of this, the publisher cannot assume any responsibility for the accuracy of this information. Each user is urged, at their own risk, to verify carefully whether the amounts of substances, experimental protocols or other types of information are plausible based on the understanding of a natural scientist. In situations where there is any doubt at all, the reader is emphatically advised to consult with a knowledgeable colleague; even the publisher is willing to offer support in clarifying any potential case of doubt. In spite of this, every application described in this volume must be performed at the user's own risk.

Introduction

Analytical chemistry, often shortened to (chemical) analysis, with its extremely varied methods, plays an important role in almost all aspects of our 'material' life, and its significance is still increasing. In spite of its recognized importance in universities in Germany, this specialized area of chemistry still has not been given the recognition it deserves by having its own professorships. Since the 1970s, analytical chemistry has developed from a classical, largely chemically oriented discipline to a physical-chemical, instrumental and problem-oriented methodology. Owing to the wide variety of methods available, it has become possible to penetrate into smaller and smaller ranges of concentrations (trace analysis) and to analyze thoroughly more complex material mixtures (e.g. using high-power separation methods). The apparatus industry which came into existence as a result of these developments represents a significant economic factor.

At the end of the 1970s, as an instructor for analytical chemistry at the University of Siegen, I and my analytical colleagues H. Monien and E. Hohaus described the importance of modern trace analysis as follows: 'Today, trace analysis data are the basis for political, legal and medical decisions, and they apply not only to the regeneration and maintenance of the quality of air, water or foodstuffs, but rather to the "quality of life," which is so often referred to. In this regard, analysis as applied to the protection of the environment can be mentioned in the same breath as air pollution control, water analysis (including oceanography) and nutritional chemistry. In medical fields, biochemical analysis and pharmaceutical research in particular rely on trace analytical methods. In pure material research and in technical fields such as materials sciences, knowledge about the levels of trace elements is an important requirement for determining the physical properties of materials. Even such diverse sciences as geology and archeology use trace analytical methods to find solutions to problems in these fields. In fact, there is hardly a field of the experimental natural sciences which does not deal with the analysis of trace elements in some fashion.'

These comments regarding trace element analysis are basically still true in the 1990s, and even with a much broader scope. Content analysis and structural analysis, i.e. the determination of concentrations and the identification and structural characterization of materials, are the two aspects or focal points of an inclusive analytical procedure.

My goal was to make clear the problems and material orientation as well as the interdisciplinary nature of chemical analysis when using physical and biochemical research methods, and to do so in both the methodological and the practical sections. Thus, an inclusive representation of chemical analysis includes statistical foundations, ideas for analytical strategies and special procedures for the analysis of water, processes or elemental species, with graphical depictions starting with sampling procedures and onwards up to the evaluation (interpretation) of the results.

1 Basic Principles

1.1 Goals of Chemical Analysis

Definition and introduction. 'Chemical analysis is the science of the extraction and use-oriented interpretation of information about material systems with the help of scientific methods' (definition from the Analytical Chemistry Section of the Society of German Chemists, GDCh).

In classical chemistry, the only goal of analysis (analytical chemistry) was to determine the composition of materials and mixtures. Today, on the one hand, it provides a service function which extends far beyond chemistry itself to almost all areas of the sciences, of medicine and of technology, even to the social sciences (archaeology, manuscript illumination, etc.). On the other hand, chemical analysis represents an independent subdiscipline of chemistry, which is closely related to physics, to measuring technology and to the information sciences. If applied meaningfully it requires interdisciplinary cooperation between the analytical chemist (analyst) and experts from the fields previously mentioned. Results of chemical analyses lead to technological, medical and legal decisions (e.g. dealing with the environment), which also demonstrates the great responsibility placed on the analyst.

A. Content analysis. Qualitative analysis (1) determines the kind of a material (e.g. a chemical element) along with information about the smallest detectable concentration. Quantitative analysis (2) determines the concentration (or amount) in a material system (the matrix). This kind of analysis is called content analysis. If one assigns z to the type of the component (material) and y to the amount of these components, then for elemental analysis (elements in this connection include atoms, ions, radicals and molecules) in the form of the qualitative analysis we arrive at a one-dimensional representation of the nature of the components. The result of a quantitative analysis can be depicted in either one or two dimensions. The two-dimensional representation, i.e. the separate diagram of the concentration of the material (3), corresponds to a chromatogram, a spectrum or a polarogram in instrumental analysis.

B. Distribution analysis. If, in addition to z (component) and y (amount), the analytical information also includes spatial coordinates (l_x, l_y, z_1, z_2, z_3), then this is a distribution analysis (1), with different materials on a surface. For example, with the help of laser microspectral analysis, surface layers can be scanned point by point. For each point, statements are made relative to the type (z) and amount (y). The result is a quantitative profile (2). These kinds of analytical methods are becoming increasingly important for determining inhomogeneities on and in solids, for example in the manufacture of microchips.

C. Process analysis. Besides the variables z and y, time t is included in place of the spatial coordinates l (1, three-dimensional; 2, two-dimensional representation) of process analysis. Process analyses, which are also called dynamic analyses, serve to monitor and control procedural operations in the chemical and pharmaceutical industries. The execution of process analyses necessitates analytical methods with sufficiently short analysis times (up to about 10 min).

D. Structural analysis. The determination of the ordering and connection of elemental components (atoms and functional groups) in molecules and solids leads to elucidation of the structure. In principle, structural analysis represents distribution analysis at an atomic level. Qualitative structural analysis provides information about the constitution, whereas quantitative structural analysis makes it possible to determine the configuration and conformation of molecules or of the elemental cells of crystals, with the help of molecular spectroscopic (MS, mass; H-NMR, nuclear magnetic resonance methods; IR, infrared spectrometry) and diffraction methods.

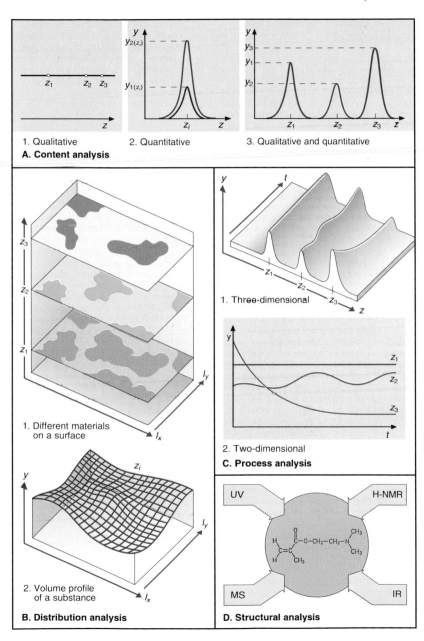

1. Qualitative
A. Content analysis

2. Quantitative

3. Qualitative and quantitative

1. Different materials on a surface

2. Volume profile of a substance

B. Distribution analysis

1. Three-dimensional

2. Two-dimensional
C. Process analysis

UV   H-NMR

MS   IR

D. Structural analysis

1.2 Formulation of the Problem and Analytical Strategy

Introduction. In order to be able to plan out an analytical strategy, i.e. the process from sampling to acquiring and processing the analytical information, first it is necessary to formulate the problem description. Once the object of the study has been characterized, the materials to be analyzed have been fixed in the prescribed matrix and the expected concentration range has been narrowed down, this is the point where the sampling is done, depending on the type of object being examined. Naturally, the sampling will be different for the determination of metal levels in an ore sample than for the analysis of heavy trace metals in the liver of an animal.

Sample preparation before analysis. The sample preparation might consist of a decomposition in which a liquid is made from the entire solid material using an acid mixture; or it might include concentration enrichment of dissolved trace materials from a water sample in order to make the materials detectable for a given analytical method; or it might require the conversion of a material into a detectable form, such as the volatilization of elements as hydrides for subsequent atomic spectroscopic analysis. Sample preparation is also often associated with the separation of interfering matrix components before a special method can even be used. On the other hand, it might be necessary to separate mixtures for the subsequent analysis of the individual components. Often, the separation processes are based on physical separation methods where the materials themselves are not changed. Precipitations and liquid–liquid extraction are labeled as chemical separation methods. Direct methods are those methods of instrumental analysis which make it possible to utilize a sample without any sample preparation as described.

Measuring object and methodology. The sample preparation and the use of separation methods yield the actual test object, for example a solution in which a concentration analysis or a conformational analysis can be performed or in which a material can be identified with selective physical or chemical methods. Physical methods include all techniques of spectroscopic analysis (atomic and molecular spectroscopy or spectrometry). Examples of chemical selection are complexometric titrations in homogeneous solution and other volumetric procedures. Finally, these determination methods yield an experimental result.

Measuring result—analytical result—analytical strategy. The measuring result is used to determine the composition of the test object, the presence and the concentration of one or more materials (qualitative and quantitative) or the constitution of a material. Part of the analytical strategy also includes the critical consideration of the results, error analysis, processing of the data and documentation of the results. Finally, the analyst must draw conclusions from the results obtained in this manner, with reference to the initial problem description; in other words, the analyst must perform a problem-oriented evaluation. Herein lies the special responsibility of the analytical chemist. Because of the vastly different test objects and problem descriptions, this requires interdisciplinary cooperation. As a rule, complex matrices or differentiated problem descriptions require a combination of methods, the linking of procedural steps: they form the actual teaching material of modern problem-oriented and practical chemical analysis. Even today, in spite of efficient equipment for instrumental analysis, only rarely is a direct path from the object to the analytical result possible. This places the burden on the analyst to develop these kinds of analytical strategies.

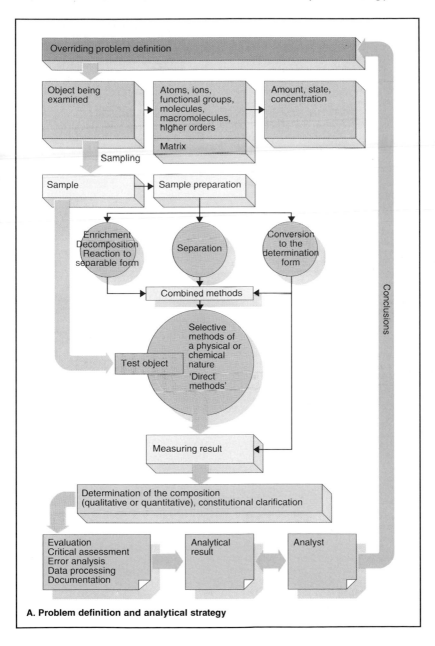

A. Problem definition and analytical strategy

1.3 Classification of Analytical Methods

Introduction. Natural laws serve as the basis of an analytical measuring principle; these include the absorption of specific wavelengths of light by defined chemical particles, for example. The analytical process as a whole consists of a series of intermediate steps. This is how information about the object to be examined and its properties with respect to a prescribed problem description is obtained.

A. Analytical principle, analytical method, analytical procedure. The analytical principle includes interactions, such as those between light of a particular wavelength and the sample, which lead to measured values which can be interpreted. The analytical principle can be described quantitatively as a substep of the measurement via the natural laws upon which it is based. Furthermore, an analytical method also includes parts of the intermediate steps 'Sample preparation' and 'Evaluation:' it represents the strategic concept for acquiring optimal information about the object being examined or measured given an analytical principle.

An analytical procedure is characterized by the procedural protocol, which includes instructions about the sampling, the sample preparation (with the required equipment and chemicals noted), about the equipment settings, the analytical calibration function, the area of application with information about selectivity, about possible errors (systematic or random errors) and about the amount of time required.

B. Classification of analytical methods. Simplistically speaking, quantitative analysis can be divided into classical wet chemical methods and (modern) instrumental methods. Historical analytical methods which to this day have retained their value as being simple, yet reliable, procedures include gravimetry (analysis by weight) and volumetry or titrimetry. The instrumental methods require special analytical techniques or equipment above and beyond scales and burettes, and often include the use of computers. They can be divided into three main groups, as follows.

The analytical principles of emission and absorption are the basis for optical methods. The interactions between atoms, molecules, ions and electromagnetic radiation generate information which can be evaluated analytically. In a narrower sense, these methods are grouped together as spectroscopic methods (atomic or molecular spectroscopic methods), either as emission or absorption spectroscopy or spectrometry. Atomic spectrometric methods include atomic absorption spectrometry (AAS), atomic emission spectrometry, e.g. in the form of flame photometry, and X-ray fluorescence analysis as an additional emission method. Molecular spectroscopic methods include such methods as spectrophotometry, infrared spectrometry (IR), mass spectrometry (MS) and nuclear magnetic reson-ance spectroscopy (NMR). They are used especially in areas of structural analysis, often also in connection (in combination or with direct coupling) with separation methods such as gas chromatography (GC).

Separation methods include chromatographic separation methods, which always require a detection method—usually one of the spectrometric methods—as a supplement to a complete analytical procedure. Generally, separation methods represent all physical-chemical divisions between two different phases, such as liquid–liquid or solid–liquid extraction and ion exchange.

Electroanalytical methods use electric current (the parameters of amperage and voltage or potential) to generate analytical information. They often include a material conversion and therefore separation processes which take place when electrons participate on the surface of electrodes, such as in polarography. On the other hand, currentless methods represent a segment of electrochemical analysis, such as potentiometry with the direct application of electrodes (as in direct potentiometry) or as an indication method in titration processes.

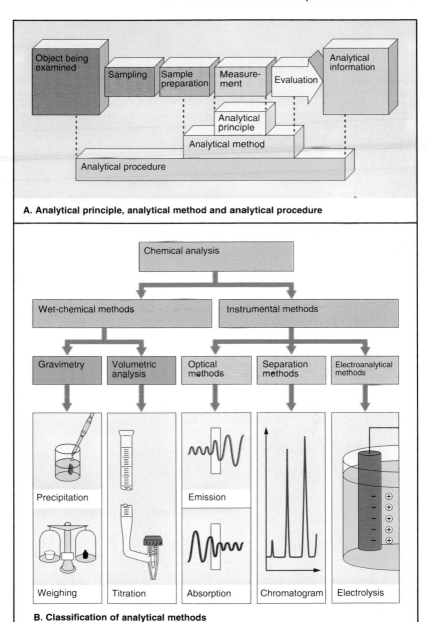

A. Analytical principle, analytical method and analytical procedure

B. Classification of analytical methods

1.4 Fundamental Working Steps and Methods in Synthesis

Introduction. Symbols are often used to represent procedural stages or work flows; these symbols provide a visual overview of how the individual steps are linked in the form of a schematic flow chart. All analytical steps are depicted by their basic principle, and not by apparatus, since the latter is continually being updated in terms of its construction and therefore any symbol representing the apparatus would quickly become outdated. The symbols shown on the opposite page, which have been incorporated into a DIN regulation, can be used to convert every analytical step into a schematic flow chart to which details about the concentration of solutions, temperature, pH, etc., can be added.

A. Fundamental procedural steps. Grinding, mixing and dissolution are the most important intermediate steps in sample preparation. An extremely wide variety of mechanical devices, such as pebble mills for finely grinding hard materials, are used for this. Mixing includes the homogenization of samples. Stirring, heating, boiling and cooling are associated with dissolution. Crystallization is less commonly used in chemical analysis; it is used more often in preparations. The addition of liquids and neutralization are part of titrimetry (volumetric analysis), whereas precipitation, heat drying, roasting, chemical drying (in a desiccator) and weighing are associated with gravimetry.

B. Separations and analytical methods. The following steps involve separation procedures or methods. Sieving involves methods for separating solid mixtures based on their particle size (sieve analysis). In filtration, solid particles are separated from liquids or gases using suitable filters (filter-paper, membrane filters, glass filter devices or porcelain crucibles for solid–liquid systems). Centrifugation involves the separation of mixtures using centrifugal force. Extraction and shaking out are two related processes: extraction describes the process for dissolving particular components out of solid or liquid mixtures. We can differentiate between solid–liquid and liquid–liquid extraction; the latter is also called shaking out.

Distillation involves the vaporization of a liquid, which can consist of numerous individual components, with subsequent condensation of the steam which is formed as a distillate (e.g. fractional distillation). Pumping gases can be considered to be a separation process in the broadest sense of the term. In a more restricted sense, electrolysis represents an electrochemical separation method; in this procedure there is a separation of ions in the form of metals, for example, on the surface of an electrode, as a result of the current flowing through an electrolyte, depending on how much current is applied.

Chromatography is a physical-chemical separation method in which a material separation takes place based on differing distributions between a stationary and a mobile phase. Depending on the structure or combination of the phase pair, one differentiates gas chromatography (mobile phase, gas) from liquid chromatography (mobile phase, liquid; stationary phase, solid or liquid) such as thin-layer chromatography (TLC; stationary phase, thin layer on a carrier substance), high-performance liquid chromatography (HPLC: stationary solid phase in a thin column) and ion-exchange chromatography.

For atomic and molecular spectroscopic methodologies, which are based on interactions between electromagnetic radiation and chemical substances, a basic symbol (spectrometry) is used with the abbreviations for the specific methods, such as NMR for nuclear magnetic resonance spectrometry, MS for mass spectrometry and IR for infrared spectrometry.

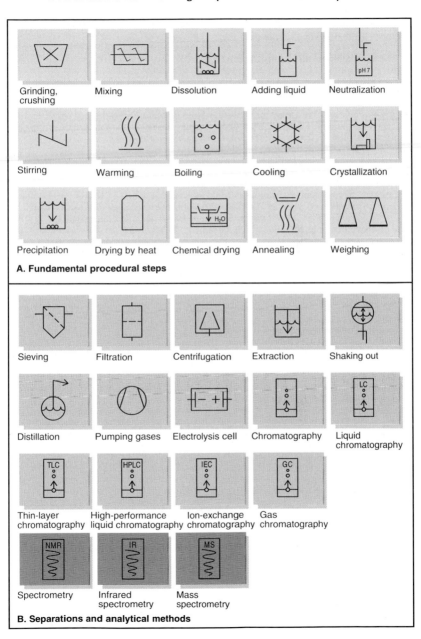

Grinding, crushing

Mixing

Dissolution

Adding liquid

Neutralization

Stirring

Warming

Boiling

Cooling

Crystallization

Precipitation

Drying by heat

Chemical drying

Annealing

Weighing

A. Fundamental procedural steps

Sieving

Filtration

Centrifugation

Extraction

Shaking out

Distillation

Pumping gases

Electrolysis cell

Chromatography

Liquid chromatography

Thin-layer chromatography

High-performance liquid chromatography

Ion-exchange chromatography

Gas chromatography

Spectrometry

Infrared spectrometry

Mass spectrometry

B. Separations and analytical methods

1.5 Fields of Application and Comparison of Analytical Methods

A. Fields of application. Limiting factors for the selection of an optimal analytical method are the amount of sample available for analysis and the expected concentration range of the material(s) to be analyzed. Since 1979, there has been IUPAC (International Union of Pure and Applied Chemistry, Oxford) nomenclature for analytical applications; in this nomenclature, three significant, associated values are defined. The mass range S ($S = m_x + m_y$) of the sample represents the range of the sample amount of component x—the analyte—in matrix y (the main component of the sample or the sum of the other components) which is necessary for a particular analytical method. The amounts of sample most commonly available are in the range of grams to upper micrograms at the least. Based on the amount of sample available, the labels macro-, meso- (or semimicro-), micro-, submicro- and ultramicro-sample are used. The label p is used to describe the exponent of the mass number 10^P. It can also be multiplied with milliliters (mL) or moles (mol), except with the unit grams (g).

The term absolute mass range Q describes the amount of analyte x for which an analytical procedure is to be applied in order to quantify it. The concentration range C is the ratio of the mass of the analyte $x(m_x)$ to the sum of m_x and the mass of the matrix m_y, or the total amount of sample. Thus, two of the defined values sample mass range, absolute mass range and concentration range determine the limits of the third value.

Depending on the concentration range (in g/g), relative to the analyte one differentiates between major components from 100 to 10% (1 to 0.01 g/g), minor components from 10 to 0.1% (10^{-2} to 10^{-4} g/g) and trace components under 0.1%. The trace component range is in turn subdivided into micro-, nano- and pico-amounts. To convert the concentration ranges C into %, ppm or ppb, multiply the exponent p by 10^2, 10^6 or 10^9, respectively. The abbreviation ppm means parts per million ($1 : 10^6$)

(1 ppm $= 10^{-4}\% = 1$ mg/kg $= 1$ μg/g), ppb stands for parts per billion (American billion, corresponding to a British milliard; $1 : 10^9$,   1 ppb $= 10^{-7}\% = 1$ μg/kg $= 1$ ng/g) and ppt means parts per trillion (an American trillion or British billion; $1 : 10^9$, 1 ppt $= 10^{-10}\% = 1$ ng/kg $= 1$ pg/g). Instrumental trace analysis makes it possible to venture into the ppt range, and in some cases even below this (ppq $= 1$ pg/kg $= 1$ fg/g).

B. Comparison of Analytical Methods. The classical methods gravimetry, electrogravimetry and titrimetry reach concentration ranges of 10^{-2} to 10^{-4} g/L. Since solutions are necessary for measurement in most cases, the quantitative unit g/L is used here. This can be used to convert to the concentration in solid samples (taking the initial weight into consideration).

Electrochemical methods such as potentiometry allow analyses up to microgram quantities per liter, or with methods such as voltammetry they extend into the micro-trace range. Photometry and fluorimetry complement one another with respect to sensitivity, but fluorimetry detects quantities about three powers of ten smaller. Atomic spectrometry has an efficiency comparable to that of chromatography, but it has its place in elemental (metal) analysis, whereas chromatography is predominantly used in the analysis of organic substances. With chromatography one should note that it is the combination of the chromatographic separation method with a detection method that results in a complete analytical process for quantitative analyses as well. Voltammetry, fluorimetry, atomic spectrometry and chromatography are characterized as especially powerful methods. Atomic spectrometry and chromatography also stand out because it is possible to analyze simultaneously many materials (inorganic or organic) using a single analytical process. For this reason, atomic spectrometry is also called a multi-element method.

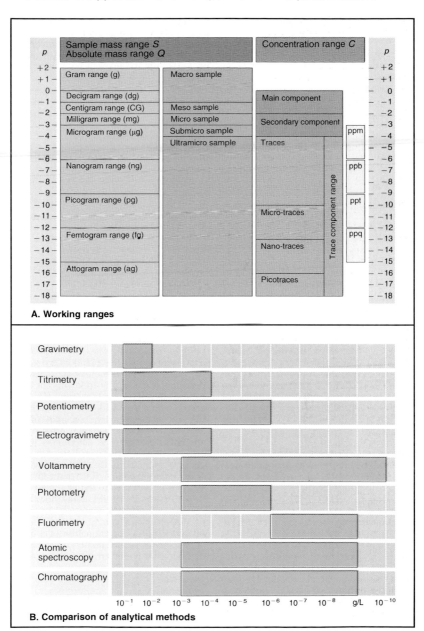

A. Working ranges

B. Comparison of analytical methods

1.6 Composite Methods and Sources of Errors

A. Composite methods and direct procedures. Composite method is the name given to the combination of methods and techniques for sample preparation, for decomposing the sample and for separating out interfering matrix components with the actual analytical method itself into a single analytical process. On the other hand, a direct procedure is one whereby the sample can be analyzed directly without any intermediate steps with the aid of non-destructive methods such as X-ray fluorescence analysis or atomic absorption spectrometry of solid matter.

As a rule, instrumental direct analysis methods are matrix-dependent relative methods. Only in rare cases is it possible to correct mathematically for the elemental cross-disturbances. For this reason, reference materials are needed to compensate for systematic errors; these materials must be very similar in their composition to the sample to be studied. However, usually an optimal sensitivity and an optimal confidence in the results can only be achieved if the trace analyte for activating the analytical signal in an isolated form is present in the smallest possible volume. For this reason, in most cases it is necessary to use the more laborious composite methods or multi-step procedures for elemental trace analysis (including trace analysis of organic materials).

B. Sources of systematic errors. Systematic errors within the individual analytical steps result in blank values or a loss of the material to be analyzed. Every intermediate step, from hydrolysis or dissolution of the analytical substrate to separation and enrichment procedures, and then to the introduction of the prepared sample in a measurable condition into the analytical device, requires reactions with an extremely wide variety of chemicals and in several containers. Since absolutely pure chemicals do not exist, and since there is no such thing as an absolutely clean container, with each intermediate step the original trace level of the analyte is increased because of the chemicals. Depending on the purity of the chemicals, this trace level that is introduced, or the blank value, can falsify the end result such that with trace analysis in the ppb range, a relative error of several hundred per cent can arise. Conversely, adsorption effects and volatilization can also have a demonstrable effect.

C. Composite methods and sources of error. Each step within a composite method can be associated with a systematic error. Depending on the step, blind values, non-homogeneity of the sample, changes in the sample and errors in weighing, measuring and calibration are the most important systematic errors one could mention. However, it is not possible in every case to differentiate strictly between systematic and random errors (variable errors, e.g. as a result of non-homogeneities in a sample).

D. Random errors. A cause can be found for systematic errors, even if this is often true only after considerable effort has been expended. In addition, random errors occur in every analytical procedure, and the sum of these errors determines the reproducibility or precision. With a homogeneous analytical sample, these random errors can be determined for separation steps and for determination steps by following the procedure shown in the diagram. Random errors cannot be avoided; they determine the precision but not the accuracy of the analytical result. The partial error of an analytical procedure is determined using error resolution: after dissolving a homogenized sample, six divided samples are removed by aliquoting. Each of the divided samples is subjected to the first analytical step (separation in this case), then one divides each of these samples into exactly two halves, each of which is subjected to the second substep (determination of a component in this case). Statistics can be used to determine the partial error from the results. If systematic errors occur, the true value lies somewhere outside of the range indicated by the random error.

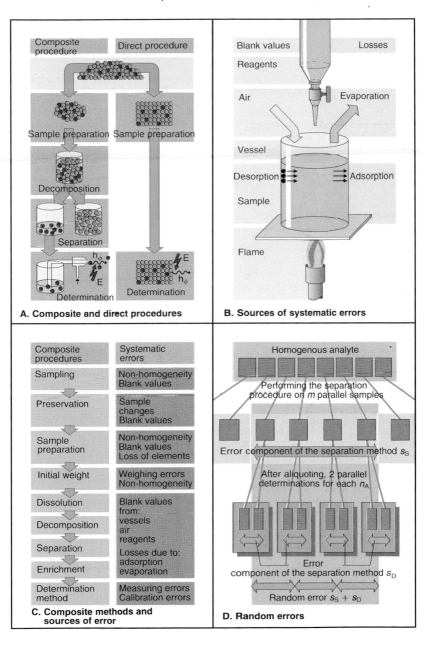

A. Composite and direct procedures

Composite procedure | Direct procedure

Sample preparation · Sample preparation

Decomposition

Separation

Determination · Determination

B. Sources of systematic errors

Blank values — Losses

Reagents

Air — Evaporation

Vessel

Desorption — Adsorption

Sample

Flame

C. Composite methods and sources of error

Composite procedures	Systematic errors
Sampling	Non-homogeneity Blank values
Preservation	Sample changes Blank values
Sample preparation	Non-homogeneity Blank values Loss of elements
Initial weight	Weighing errors Non-homogeneity
Dissolution	Blank values from: vessels air reagents
Decomposition	
Separation	Losses due to: adsorption evaporation
Enrichment	
Determination method	Measuring errors Calibration errors

D. Random errors

Homogenous analyte

Performing the separation procedure on m parallel samples

Error component of the separation method s_S

After aliquoting, 2 parallel determinations for each n_A

Error component of the separation method s_D

Random error $s_S + s_D$

1.7 Statistical Evaluation of Analytical Results

A. Calculating averages. One of the tasks of the analyst is to evaluate the results of the analysis or measurement. From a limited number of measurements (e.g. $n = 20$), the arithmetic mean is derived using $\bar{x} = (\sum x_i)/n$ (a small number of measurements: random sampling). The term population represents an unlimited number of measurements using the same material. The subset average x is an estimate of the arithmetic mean μ of the entire population. If there are no systematic errors, this value will be very close to the actual value. Values can be located very close to one another, with little scatter around the mean value, or they can have a great deal of scatter and still result in the same mean value. The only difference between the two sets of numbers is in the size of the standard deviation, i.e. the precision.

B. Normal distribution (Gaussian curve). The precision, or reproducibility, of a procedure is given as the standard deviation s, which is the square root of the mean error squared:

$$s = \sqrt{\frac{1}{n-1} \sum_{i=1}^{n} (x_i - \bar{x})^2}$$

The standard deviation has the same dimensions as the mean value. The relative standard deviation is defined as the ratio of the mean value and the standard deviation (and is usually expressed as a percentage). The smaller the standard deviation (filled circles in A), the greater is the precision. If there are a large number of measurements, the value x can be displayed as a function of the frequency of occurrence, often in the form of a Gaussian or normal distribution. The maximum of the curve, at the point where $\bar{x} = \mu$, is derived from the arithmetic mean of all measured values. The width of the curve, shown as the distance between the two inflectional tangents, represents the standard deviation σ (random error). Some 68% of all values lie in this range of $\bar{x} \pm \sigma$. The probability (statistical certainty) is greater than 95% that a measured value with a random error lies within the range $\bar{x} \pm 2\sigma$. The interval is called the confidence interval. The most important properties of the curve consist of the symmetrical (normal) shape, where for each positive value there is a negative value of the same size, and in the relative frequency of values with a small deviation from σ.

On the other hand, the relative frequency of values with a large deviation from the mean value is small: only 0.26% of the measured values lie outside $\bar{x} \pm 3\sigma$.

C. Blank values, limits of detection and determination. The smallest value which can be detected with sufficient statistical certainty depends on the sensitivity of the procedure, the blank value and its scatter. The value w_N at the detection limit (diagram 1) defines the interference level ('noise level' when working with an electronic measuring device), with at least three blank value standard deviations s_b above the blank value x_b. However, owing to the scatter of the blank value and the measured values, measured signals are only received 50% of the time by analytical values at the detection limit; the other 50% are blank values. Half of the area of the Gaussian curve 2 lies within the shaded interference level (1). A value is different from the interference level with 99.7% certainty (P) if it can be described by $x_E = x_b + 6s_b$ (see diagram 2). The term w_E is the measured value and y_E is the analytical value at the limit of detection, which represents the smallest amount of material that can be significantly differentiated from zero. The three curves show the measured value distribution of the blank values and of the measured values at the detection and identification limits. An example of photometric analysis is shown in diagram 2.

D. Systematic and statistical errors. Random errors lead to positive and negative deviations from the true value; systematic errors produce variations in just one direction. The figures showing the scatter of shots at a target provide a visual representation of the differences between statistical (random) and systematic errors.

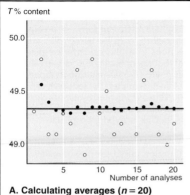

A. Calculating averages ($n = 20$)

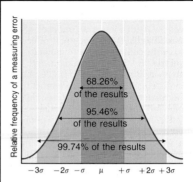

B. Normal distribution (Gaussian curve)

68.26% of the results

95.46% of the results

99.74% of the results

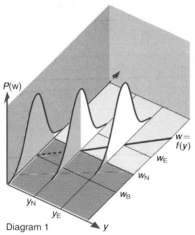

Diagram 1

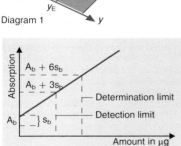

2. Example of a photometric determination

C. Blank values, limits of detection and determination

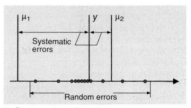

1. Systematic and random errors

2. Statistical error: small; systematic error: 0

3. Statistical error: large; systematic error: 0

4. Statistical error: small; systematic error: large

5. Statistical error: large; systematic error: large

D. Systematic and statistical errors

E. Calibration curves having different slopes. In the case of linear calibration curves, i.e. with a linear dependence of the measured signal on the analytical value, the slope of the calibration curve indicates the sensitivity E. Thus, process a has the greatest sensitivity if the example shows three different analytical procedures for a single material (e.g. comparing polarography, photometry and titrimetry). The coefficients of the individual calibration functions indicate the partial sensitivities with which the separate (physical) measured values correspond to the change of the analytical value (amount, concentration) of the different components (a, b, c).

F. Calibration curve for riboflavin. The quantitative analysis of the vitamin riboflavin in a substance such as a foodstuff can be done using fluorimetry. In this case, the abscissa shows the ppm value (relative to the sample used), and the ordinate shows the measured relative fluorescence intensity (shown here as a percentage of the device's deflection). As a rule, calibration functions are determined for at least a power of ten of the concentration (as the working range). In fluorimetry, smaller linear regions are often determined.

G. Falsification of calibration curves. Sampling and sample preparation are significant sources of error in an analytical procedure. Systematic errors are generated by matrix effects. Additional causes are incorrect calibration due to unsuitable calibrated test pieces and methods. Based on their influence on the quantity to be measured W, we differentiate between additive (constant), multiplicative (linearly proportional to the measured value), and non-linear measured value-dependent errors. Additive errors (curve 2), perhaps due to unrecognized blank values, generate a displacement parallel to the true calibration curve (1). Multiplicative errors (matrix effects) change the slope of the calibration curve, usually producing a smaller slope (curve 3). With non-linear value-dependent errors, both the spatial position and the form of the calibration curve (4) are changed.

H. Gravimetry of phosphate. When determining phosphorus gravimetrically as phosphate, different forms of precipitation or weighing can be used: after chemical reactions, phosphorus can be weighed out as molybdophosphate (E_1), silver thallium phosphate (E_2) or magnesium diphosphate (E_3). In gravimetry, the sensitivity (E) of the procedure is indicated by the stoichiometric factor. Thus, the greatest final weight will be obtained as molybdophosphate.

I. Calibration curve for $A = kc + b$. This calibration curve for a photometric determination shows a linear course, the size of the blank value b being based on the intersection with the ordinate and the scatter of the individual calibration values. If the blank value is subtracted from the measured values, or if measurements of the sample vs the blank solution are made, the line moves through the zero point.

J. Calibration curve for iron. This shows a calibration curve for a photometric determination whereby the measured values are given in extinction (absorption) units, and the concentration is 10^{-5} mol/L. The ratio $\Delta A/\Delta c$ includes the sensitivity E of this method.

K. Bowed calibration curve. Copper analysis is used as an example of a bowed calibration curve in which a linear region up to approximately 5 mmol can be seen; this can be used as a working region.

L. Standard addition method. To be able to recognize matrix effects, frequently calibrations are performed by adding known quantities of the substance to be analyzed. The concentration after this addition is plotted on the abscissa, and the zero value corresponds to the sample's measured value. The intersection of the dotted line with the abscissa yields the concentration in the sample (with a negative sign in this case).

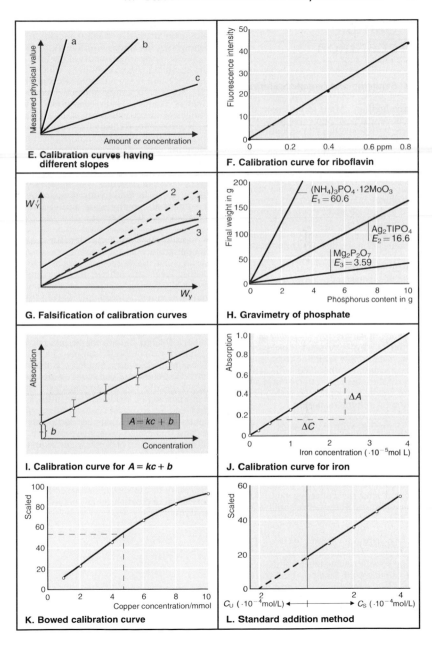

E. Calibration curves having different slopes

F. Calibration curve for riboflavin

G. Falsification of calibration curves

H. Gravimetry of phosphate

I. Calibration curve for $A = kc + b$

J. Calibration curve for iron

K. Bowed calibration curve

L. Standard addition method

2 Sample Preparation

2.1 Sampling and Sample Stabilization

Introduction. After formulation of the problem and development of an analytical strategy (see Section 1.2), sampling represents the first significant, result-determining step in every analysis.

A. Degree of error in sampling. The main goal in sampling is to obtain and maintain (stabilize) samples which are representative and which make possible a problem-oriented statement from the results after the complete analysis, based on the sampling system and procedure. As a rule, the degree of error in the sampling portion of the analytical procedure is considerably higher than that in the methodology itself.

B. Sampling gases. Representative sampling of gases is often particularly difficult because gas samples can contain not only gaseous materials, but also materials in a liquid (aerosols) or solid (dust particles) state. One simple sampling technique is to use a sampling tube filled with activated charcoal for the adsorption of organic solvent vapors in conjunction with a bellows pump for volumetric dosing.

C. Sampling with glass vessels. The direct sampling of gaseous materials is also possible using glass vessel or special plastic containers, if effects from adsorption on the walls are negligible. The sample is drawn in through the glass tubing using a pump located after the collecting tank. After a predetermined purge period, both sides are closed.

D. Absorption from gases. Liquid absorption systems make selective sampling possible owing to the ability to select the solution used. Small gas bubbles, which are essential for effective absorption, can be obtained using impinger gas bottles or glass frits.

E. Sampling from water. Random samples taken manually or automatically from pipes, bodies of water and effluents provide a picture of the water quality at the time of the sampling. Mixed samples, which are usually taken using an automatic sampler (see Section C), can be obtained based on time or volume. Time-proportional sampling is when equal volumes are taken from a stream of water at equal time intervals for a mixed sample. Flow-through proportional sampling uses changing volumes at constant time intervals, proportional to the effluent. Volume-proportional sampling takes constant volumes, but at different times, so that the total volume removed over a somewhat lengthy time period corresponds to the flow volume.

F. Automatic sampling system for water. Scooping buckets for layers close to the surface or sealable scooping apparatus for deeper levels are utilized for the manual removal of water samples. In automatic sampling systems, the sample is taken from a continuous sample flow ('conveying,' 1) following a pulsed circuit (time or volume determined), directed into a sample container ('dosing,' 2), and removed as a partial sample for a composite sample (3 and 4). The contents of water samples, e.g. nitrate, nitrite and ammonia, can change especially due to the effect of microorganisms. Other materials can be oxidized (e.g. sulfide) or they can be adsorbed on the walls of the container (trace materials, e.g. metal ions or organic materials). For this reason, it is often necessary to preserve water samples. For example, it is possible to stabilize substances such as nitrate, nitrite or ammonia by cooling them to $4\,°C$. In other cases it is necessary to use special additives for preservation, depending on the nature of the analytes.

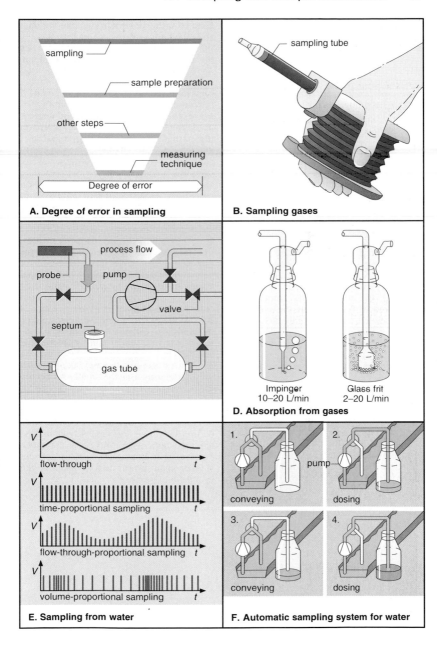

A. Degree of error in sampling

sampling

sample preparation

other steps

measuring technique

Degree of error

B. Sampling gases

sampling tube

process flow

probe

pump

valve

septum

gas tube

D. Absorption from gases

Impinger
10–20 L/min

Glass frit
2–20 L/min

E. Sampling from water

V

flow-through t

V

time-proportional sampling t

V

flow-through-proportional sampling t

V

volume-proportional sampling t

F. Automatic sampling system for water

1.
conveying

2.
pump
dosing

3.
conveying

4.
dosing

G. Sampling from solid materials. A defined number n of individual samples are taken from a population such as a pile of refuse, following a pre-determined pattern (H). Overall, they constitute the raw, mixed or total sample. The actual analytical sample is obtained from the divided sample by repeated grinding, sieving and dividing (J); this is also referred to as (sample) reduction. The requirements which this flow chart must meet are as follows: the test sample must be representative of the raw sample and the raw sample must in turn be representative of the population. If one assumes that the properties of the population can be considered to have a normal distribution (see Section 1.7), then a defined number of individual samples n (the raw sample) and an equal number of analyses will yield the characteristic mean value for a substance, and the degree to which this figure approaches the true value μ is called representativeness. Since the sampling error noticeably exceeds the analytical error as a rule (see A), in general the sampling is considered to be perfect if the sampling error does not exceed roughly 3/4 of the total error, i.e. if the condition $s_p : s_A = 1.3$ is met (s_p = sampling error, s_A = analytical error). This is valid for s as the standard deviation for property x in a random sampling:

$$s = \sqrt{s_p^2 + s_A^2}$$

H. Sampling from soils. Simple equipment such as shovels and siphon cylinders can be used here. If a plunging cylinder is used, the soil structure is retained. Since the soil sample is supposed to be representative of the total area to be analyzed, a variety of removal techniques can be suggested. If there are spatial inhomogeneities, the normal method is used. With relatively homogeneous soils, the sampling surface can be reduced in the form of parcels, diagonals or cross strips.

I. Minimum mass and particle size. The sampling error increases with decreasing content of one of the components to be analyzed in the heterogeneous mixture, with decreasing sample mass or with a corresponding decreasing sample volume, and especially with increasing particle size. For a graphical representation, if one uses the edge length a of a particle (volume $V = a^3$), one obtains a line which can be used to estimate the sampling error in a double logarithmic coordinate system. In practical operation, there are parallel families of curves for the necessary minimum mass of an average sample which are dependent on the particle size of the largest particle, whereby a–f are labeled as homogeneity parameters. Owing to this dependence, sampling is usually linked with a screening analysis (see also Section 1.4). Based on guidelines for the removal of samples from refuse and deposited materials set up by the German State Working Group for Refuse, the minimum mass of a sample is given as m_{min} (kg) = 0.06 a (mm). However, using this rule of thumb, considerably smaller sample masses are required than in Taggart's graph; of necessity, this results in considerably higher sampling errors (see A).

J. Sample splitting. It is possible to reduce the amount of a sample (e.g. with ore loading) using sample reduction, by splitting the sample into quarters or by using a mechanical sample divider if the particles are sufficiently fine or small. A sample splitter is a bent tube, open at one end, in which the total sample is continually divided in half by rotating the tube. If soil samples need to be stored for some time before the actual analysis, they should be air-dried and crushed and stones should be removed with a fine sieve (with a mesh size smaller than 2 mm) before packing the sample in plastic bags.

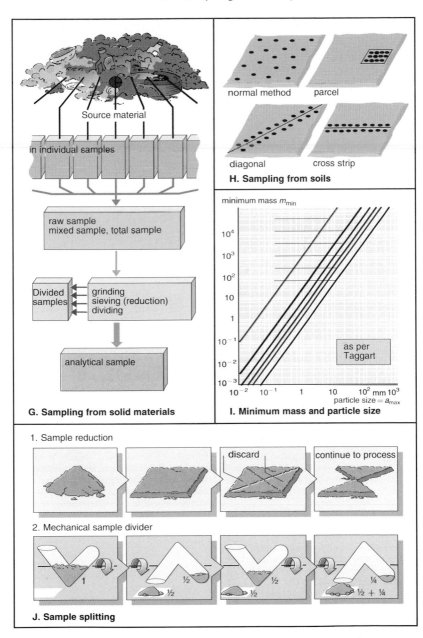

Source material

in individual samples

raw sample
mixed sample, total sample

Divided samples ← grinding
sieving (reduction)
dividing

analytical sample

G. Sampling from solid materials

normal method   parcel

diagonal   cross strip

H. Sampling from soils

minimum mass m_{min}

10^4
10^3
10^2
10
1
10^{-1}
10^{-2}
10^{-3}

as per Taggart

10^{-2} 10^{-1} 1 10 10^2 mm 10^3
particle size = a_{max}

I. Minimum mass and particle size

1. Sample reduction

discard   continue to process

2. Mechanical sample divider

1   ½   ½   ¼
½   ½   ½ + ¼

J. Sample splitting

2.2 Decomposition Methods

A. Systematic sources of error. The decomposition of a substance or a mixture of substances does not refer to the dissolution, but rather the conversion of slightly soluble substances into acid- or water-soluble (ionogenic) compounds (chemical dissolution). Therefore, a decomposition is associated with a chemical change in the starting material or the matrix as a whole. Similar to dissolution processes, decompositions generally have fairly large sources of error: material losses can occur as a result of volatilization (1a; e.g. with Hg traces) and adsorption effects (1b) in decomposition vessels. Blank value carry-overs can be traced back to the reagents used (2a; acids, alkaline solutions, etc.), to impurities in the atmosphere (2b) with open decomposition systems (via air, dust), to impurities in the vessel materials (2c; via desorption and solubilization of the vessel materials) and to a slight degree to the other objects needed, such as a spatula or glass rod (2d). The magnitudes of the effects depend largely on the decomposition procedure which has been chosen. Based on the phase status of the decomposing agent, there is a division into melt decompositions, wet decompositions (with liquid decomposing agents) and decompositions by combustion.

B. Pressure (melt) decomposition with Na_2O_2 (Parr bomb). Melt decompositions are the classical procedures of technical analysis and mineral analysis. In elemental analysis, organic substances which contain halogens or sulfur are reacted with an oxidizing agent, such as sodium peroxide, in a closed, pressure-resistant container such as a Parr bomb; analyzable inorganic salts are formed in the process.

C. Pressure decomposition container (as per Tölg). Contamination and losses with wet decompositions (e.g. in an open platinum container) can be largely avoided if the procedure is performed in a pressure bomb. Small weighed samples (200 to 500 mg) and increased temperatures which speed up the decomposition are other advantages of these closed systems. Pressure decomposition bombs as per Tölg are used especially in trace element analysis, mainly for organic and biological matrices, but also for inorganic samples and alloys and metals (decomposition with acid) and for minerals (with lye). The pressure decomposition container as per Tölg is made in principle of the following components: pressure valve, screw-cap, pressure spring, metal lid, PTFE lid, PTFE insert and a pressure vessel (lined with stainless steel and thermal insulation). It rests in an aluminum heating block with a thermal probe, cold water connection and heating current connection with an overheating fuse. During decomposition in Teflon (PTFE) vessels, the temperature limit is about 170 °C. Recently, quartz glass inserts have come into use; they allow an increase in the decomposition temperature up to 300 °C.

D. High-pressure incinerator (as per Knapp). Knapp's high-pressure incinerator makes it possible to perform decompositions at 300 °C which generate practically carbon-free solutions. By adding an external pressure of 100 bar (nitrogen), the vapor and reaction gas pressures generated during the decomposition are compensated. The autoclave with the heating element contains an autoclave lid and lock ring. The heating block insert contains the quartz decomposition vessel with its own lid, and this vessel in turn contains the sample which is to be decomposed, along with acid (nitric acid as a rule). The quartz decomposition device is closed using a thin PTFE foil and a quartz lid. The elements (metals) to be analyzed, which lie in the decomposition solution, are present exclusively in inorganic-ionic form, so that they can be analyzed universally with any method suitable for determining metals. The decomposition times in the Tölg bomb and in the high-pressure incinerator lie between 1 and 3 h.

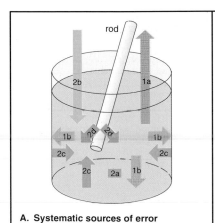

A. Systematic sources of error

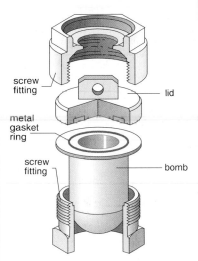

B. Pressure (melt) decomposition with Na_2O_2 (Parr bomb)

screw fitting

lid

metal gasket ring

screw fitting

bomb

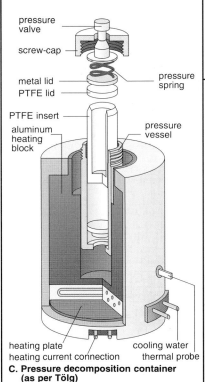

pressure valve

screw-cap

pressure spring

metal lid
PTFE lid

PTFE insert

pressure vessel

aluminum heating block

heating plate
heating current connection

cooling water
thermal probe

C. Pressure decomposition container (as per Tölg)

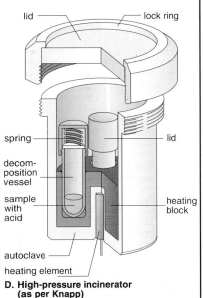

lid

lock ring

spring

lid

decom-position vessel

sample with acid

heating block

autoclave

heating element

D. High-pressure incinerator (as per Knapp)

E. Wet incineration automat. A continuous sample throughput in routine operation is possible in a commercially available wet incineration automat. The device is suitable for decompositions with hydrofluoric, sulfuric, nitric and perchloric acids, especially for organic and biological matrices, as well as for the determination of nitrogen by the Kjeldahl method with sulfuric acid decomposition. The sample material is weighed into the incinerator vessel, along with the decomposition reagents. Three or four vessels (made of borosilicate glass, quartz glass or PTFE) are hung in conveyor racks made of acid-resistant sheet steel, and the vessels are brought to the input section of the automat. A transport mechanism lifts the racks up and brings them into the heating zone, where the vessels are immersed in five successive metal heating blocks. The transport intervals can be adjusted (between 10 min and 5 h) using a timer. The immersion depth can also be adjusted, which regulates the temperature in the decomposition vessel (up to 400 °C). The heating zone consists of a hotplate on which the five aluminum heating blocks are situated. The temperature of the heating blocks decreases on going from the bottom to the top, so a temperature gradient is produced on a hot-plate which is regulated at a constant temperature. After the conveyor racks have proceeded through the heating zone, they are cooled in the output area.

F. UV decomposition device. In some elemental determination methods, such as polarography/voltammetry (see pp. 62*ff*) and photometry (see pp. 108*ff*), it is necessary to destroy the organic materials completely. Examples are heavy metal determinations in waste waters or liquid foodstuffs such as beer and wine. Oxidative decomposition with the addition of hydrogen peroxide and UV radiation is used in this case. In contrast to the wet-chemical thermal decomposition processes described previously (in an open or pressurized vessel), UV decomposition requires only small amounts of acid in addition to hydrogen peroxide, without a significant increase in temperature. This keeps the contamination resulting from heavy metal levels in the reagents at a low level. Highly reactive chemical radicals are produced by UV radiation, and ozone is generated. In the secondary reactions, organic substances which bind to heavy metals (i.e. they form complexes and therefore evade voltammetric or photometric analysis) are degraded (decomposed). The sample solutions are located in quartz vessels. Since the sample solution heats up only slightly in an open vessel (largely avoided by cooling), there is no loss of (water vapor-) volatile substances. The method of determination which is to be used will dictate how complete the decomposition of organic substances has to be. Several quartz glass reaction vessels can be arranged around the UV emitter in the center. Ozone-containing gases can be drawn away using a laboratory hood.

G. Microwave oven. Microwave ovens, which can be found in almost every household today, have been used recently in microwave decomposition systems. Instead of thermal initiation, microwave initiation is utilized here. The advantages over thermal initiation lie basically in the shorter time needed (several minutes instead of 1–3 h). The sample or decomposition solution is directly heated, but the vessel itself is only indirectly heated (by the solution as it warms up). Teflon (PTFE) pressure vessels for pressures up to 7 bar and temperatures up to 200 °C are used for this type of decomposition; discharge valves prevent the decomposition vessels from overloading. Wave reflectors or a turn-table are used to distribute the energy (up to 600 W) equally.

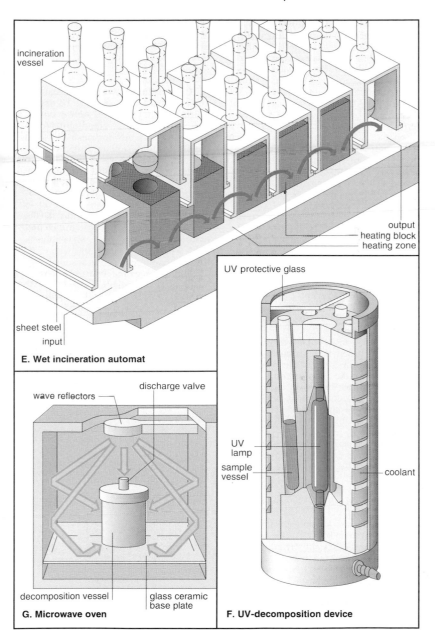

incineration
vessel

output
heating block
heating zone

sheet steel
input

E. Wet incineration automat

UV protective glass

UV
lamp

sample
vessel

coolant

wave reflectors

discharge valve

UV
lamp

decomposition vessel   glass ceramic
base plate

G. Microwave oven

F. UV-decomposition device

H. Oxygen bomb. Almost 1000 years ago, the French chemist Berthelot described the combustion of organic materials with the aid of oxygen under increased pressure in a calorimetric bomb. The powdered analyte is pressed into tablets and placed in a small cup in the holding mechanism on the bomb's lid. After screwing on the lid with the valve and applying a current, oxygen is introduced until a particular pressure is reached (e.g. 25–40 bar) and the sample is then ignited electronically using the electrodes; an iron wire touching the sample serves as the fuse. The combustion products, such as salts or oxides of metals such as Na, K, Mg and Ca, can be subsequently dissolved and analyzed.

I. Combustion in Schöniger flasks. Except for dry incineration in a platinum crucible, oxygen-induced decomposition requires special apparatus, especially if one wishes to avoid the loss of volatile substances. Numerous organic substances can undergo combustion in a flask (in this case, a Schöniger flask for microanalysis) filled with oxygen under normal pressure. The samples are wrapped in a strip of filter-paper (1). The flask, which contains an absorption solution, is filled with oxygen (2). The stopper with a holding mechanism (platinum mesh) for the slip of paper is quickly placed on the flask after ignition (3). During combustion (4), the flask is held upside down so that the absorption solution seals off the stopper. Elemental analyses (halogens, sulfur) in particular are performed using this technique. The decomposition solution can be used for subsequent titrimetric (see p. 40) or ion-chromatographic (see p. 160) analyses.

J. Combustion in a Wickbold apparatus. Gaseous, liquid and solid samples can be decomposed in a oxyhydrogen flame by the Wickbold method. The substance is vaporized in the oxygen stream from the vaporization chamber and then fed into the oxyhydrogen flame, which burns in a cooled quartz tube. The combustion products are condensed here and also in a connected cooling coil, or they are captured in an absorption solution as gaseous materials. Combustion in a Wickbold apparatus is one of the quickest and most effective methods for destroying organic materials of all types, even for the decomposition of dried sewage sludge.

K. Combustion in an oxygen stream. The quartz apparatus contains a combustion chamber in the middle. Above this is a cooling system with liquid nitrogen, and below it is a vessel in which the sample (decomposition) solution collects. Using a carrier, the sample is brought into the combustion chamber, where it is ignited in an adjustable oxygen stream using an infrared radiator.

L. Cold plasma incineration. Excited oxygen (a plasma) is generated under reduced pressure using a high-frequency field, whereby oxygen atoms with a short lifetime mainly appear. Because of the rapid recombination of the atoms, the sample is positioned in a small boat immediately after the high-frequency coil. The required low pressure is generated using a pump behind the cold trap. This way oxidation is achieved at relatively low temperature, and easily volatile materials (such as compounds of As, Sb, Se and Te) can be quantitatively detected in the process. This method is used especially for trace determinations of metals in organic (biological) materials.

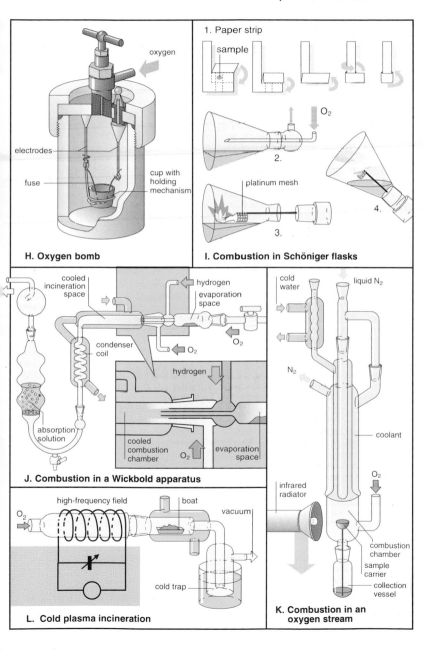

H. Oxygen bomb

I. Combustion in Schöniger flasks

1. Paper strip

J. Combustion in a Wickbold apparatus

L. Cold plasma incineration

K. Combustion in an oxygen stream

2.3 Enrichment Methods

A. Schematic representation of an enrichment procedure. The term 'enrichment' or preferably 'preconcentration' encompasses analytical procedures which are methodologically very different from one another, and which result in an increase in the concentration of the solutions, or of the mass ratio of the trace components to the matrix. Enrichment procedures are always necessary when the concentration of the chemical trace materials that are to be determined is so low that the available determination methods, which are almost exclusively apparative, are insufficient. With absolute enrichment, the trace materials are transferred from a large volume to a smaller volume of a second phase. Relative enrichments occur when there is an increase in the concentration without phase transfer, via the partial removal of the matrix (e.g. removal of the water by concentrating the solution). If the trace material to be determined is selectively separated out of the matrix with a concurrent increase in concentration into another phase, this is selective enrichment. Significant factors in the selection and evaluation of enrichment methods are the recapture of the trace materials (in per cent) and the enrichment factor (as a concentration ratio of the trace in both phases, e.g. in liquid–liquid extraction; see C) for recapture yields of less than 100%.

B. Phase conversion: liquid–gas–liquid. The systematics of trace element enrichment are oriented toward phase conversions in the course of the enrichment of matter. For example, thermal volatilization with the aid of distillation makes it possible to separate arsenic from a hydrochloric acid solution as arsenic trichloride in conjunction with a reduction in the volume of the solution after condensation.

C. Phase conversion: liquid–liquid. Liquid–liquid extraction is used very frequently: traces of heavy metals in a water sample can be transferred as metal complexes into an organic phase which is practically immiscible with water by adding a chelating agent such as dithiocarbamate; then the traces can be enriched. Even organic, slightly dissociated (non-polar) materials can be enriched in an organic solvent without the addition of reagents. In addition, the organic solvent can then be evaporated off, and the residue can be dissolved in a smaller volume of liquid.

D. Coprecipitation. By using inorganic trace element detectors such as iron or aluminum hydroxide, it is possible to coprecipitate trace elements from sea water or other matrices; in this procedure the detector is first generated in a large sample volume by precipitation. Once the precipitate has been filtered off and allowed to dry on a piece of filter-paper, it can be used directly for X-ray fluorescence analysis, for example of Mo.

E. Enrichment via ion exchange (sorption). Like coprecipitation, liquid–solid phase transition is used for enrichment: depending on the nature of the materials, ion exchangers, adsorbents such as activated charcoal and polymers, can be used for sorption from larger volumes. Usually it is necessary to follow this up with elution in a small volume of solution.

F. Enrichment by electrolysis. Another means of enrichment with liquid–solid phase conversion is electrolysis (see p. 52) on different electrode materials, including coal. Cathodic preconcentration is also a component of an analytical method in the form of inverse voltammetry (see p. 62).

G. Enrichment in cartridges. In conjunction with chromatography and atomic absorption spectrometry, trace elements are enriched on adsorbents in cartridges (which is possible even from the air), and from there they are eluted into the analytical device. This avoids contaminations in particular, and automation is achieved within the framework of a composite process.

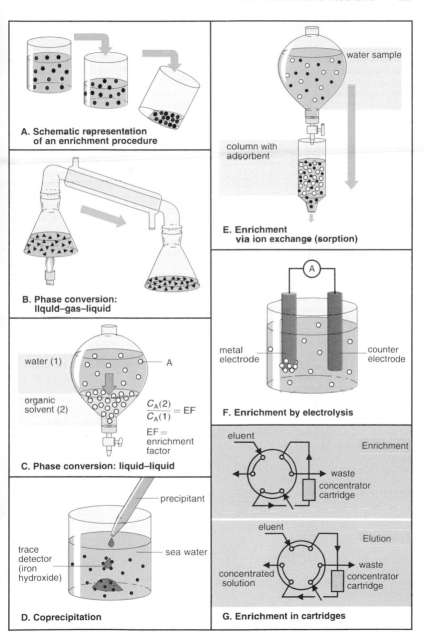

A. Schematic representation of an enrichment procedure

B. Phase conversion: liquid–gas–liquid

C. Phase conversion: liquid–liquid

water (1)

organic solvent (2)

A

$$\frac{C_A(2)}{C_A(1)} = EF$$

EF = enrichment factor

D. Coprecipitation

precipitant

trace detector (iron hydroxide)

sea water

E. Enrichment via ion exchange (sorption)

water sample

column with adsorbent

F. Enrichment by electrolysis

metal electrode

counter electrode

G. Enrichment in cartridges

eluent

Enrichment

waste

concentrator cartridge

eluent

Elution

concentrated solution

waste

concentrator cartridge

2.4 Sample Cleanup Procedures

Introduction. In addition to decomposition and enrichment, sample preparation also includes steps for separating materials, i.e. from matrix components. The purpose of these steps is to make possible trouble-free quantitative analysis with a particular method. Efficient cleanup procedures simplify the quantitative analysis and provide a significant contribution toward obtaining correct results. On the other hand, sample preparation steps need to be monitored continually; therefore, the determination of the recovery and its reproducibility are components of analytical quality control.

A. Solid-phase extraction. It is possible to separate matrix components using special adsorbents (polymer materials, silica gel and especially chemically bonded phases such as n-alkyl groups, phenyl groups and other groups bonded to silica gel) owing to the selectivity of the adsorption and the subsequent elution with mostly organic solvents. The principles of chromatography are valid here: since we are dealing with separations between a solid and a liquid phase, this technique is also referred to as solid-phase extraction. This procedure is easy to use, with a syringe as the reservoir for the solvent, suitable connections to a column with the adsorbent packed between two frits, and a shut-off valve. The selection of the adsorbent is determined by the polarity of the materials to be extracted (analyzed) and by the type of matrix (interference). Non-polar substances such as polycyclic aromatic hydrocarbons (PAHs) can be separated (and concurrently enriched) from polar substances in water on a phase with octadecyl hydrocarbon chains. Interfering impurities are selectively removed using different washing steps before the PAHs are eluted.

B. Column switching technique. One automation approach is the column switching technique, which, like solid-phase extraction, is used in conjunction with selective adsorbent. The advantage of this technique lies in the fact that sample preparation is directly coupled with liquid chromatography, usually HPLC analysis. The sample solution is fed via a sample valve into a purification column first, and then through washing steps via valve 2 (positions 1, 2, 6, 5) into the actual separation column (positions 1, 6, 2, 3, 4, 5) in association with the detector (see also Section 2.3.G, Enrichment in cartridges, on p. 28).

C. Sweep co-distillation. The sweep co-distillation method was developed to determine residues of pesticides in oils. The sample is injected on to a heatable column filled with silanized glass beads (small particles, large surface area) and glass-wool. A readily volatile solvent is introduced using a gas supply unit (with a rotameter) and nitrogen in order to transport the substances which are soluble in the solvent from the oil, which has deposited itself on the surface of the glass beads, dissolving and vaporizing into the test-tube in the water-bath. Less volatile substances which are co-transported, or which are only slightly soluble at lower temperatures, are retained in a PTFE cooling coil. This mild distillation procedure, together with a solvent such as methylene chloride, can be used to obtain clean solutions of pesticides for GC analysis.

D. TSA procedure. The thermomicro-, separation, transfer and application procedure (TSA) developed by E. Stahl is coupled with thin-layer chromatography. The sample is introduced into a glass cartridge using a capillary tube. By heating in an oven at a prescribed temperature, the volatile substances from a matrix end up directly on the TLC layer held in front of it. The thermofractionation principle is applied here.

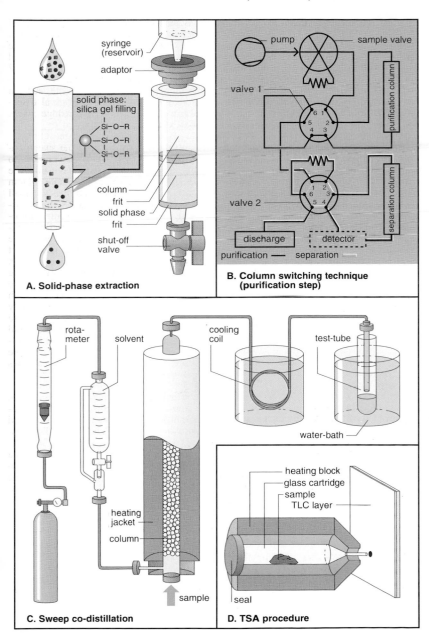

A. Solid-phase extraction

syringe (reservoir)
adaptor
solid phase: silica gel filling
$Si-O-R$
$Si-O-R$
$Si-O-R$
column
frit
solid phase
frit
shut-off valve

B. Column switching technique (purification step)

pump
sample valve
purification column
valve 1
6 1
5 2
4 3
valve 2
1 2
6 3
5 4
separation column
discharge
detector
purification — separation

C. Sweep co-distillation

rota-meter
solvent
cooling coil
test-tube
heating jacket
column
water-bath
sample

D. TSA procedure

heating block
glass cartridge
sample
TLC layer
seal

3 Detection Methods

3.1 Working Methods and Analytical Process

A. General working methods of qualitative analysis. The types of reactions which can be used analytically for classical quantitative analysis are precipitation, complex formation, neutralization, redox and gas generation reactions. Several working methods which are used repeatedly must be used in order to be able to execute procedures such as the formation of a characteristic precipitate, the dissolution of a precipitate, color changes or the development of gas as evidence for inorganic substances (usually in form of ions). These include: adding a reagent solution drop-by-drop (for all reactions types); depositing a precipitate (either via gravity in a test-tube or by using centrifugation); filtering a precipitate to separate it from the filtrate; heating a solution to speed up the reaction; and transferring a gas into a reagent solution (e.g. to detect CO_2 in a barium hydroxide solution).

Wet-chemical, single-step reactions are identification reactions. The fewer the ions that react with a particular reagent, the more selective the reagent is. If only a single ion generates a reaction, it is a specific reagent. However, this degree of specificity is achieved in only a few individual cases. In addition to these reagents, group reagents are used, such as hydrogen sulfide, ammonium sulfide and ammonium carbonate (see B). A soda extraction is performed for anion analysis: the goal is to precipitate metal ions as slightly soluble carbonates and thereby to separate them as possible interfering factors when detecting anions.

The steps involved in a qualitative analytical procedure include the following. After characterizing the substance to be analyzed as to type, amount, aggregate condition, color and odor, preliminary tests are conducted. The pre-tests include checking the flame coloration (see emission spectrometry), heating the original substance in an ignition tube, selective detection reactions such as for arsenic (Marsh test), color reactions in a borax or phosphorus salt bead, the etching test for detecting fluorides, oxidation melts for detecting chromium (as chromate) or manganese (as manganate) and the flame test for tin. Following these preliminary tests comes the detection of anions, either directly in the original material as with carbonate or in the filtrate of the soda extract. This is followed by cation analysis, for which the sample must first be dissolved. If after dissolving it in water, hydrochloric acid and nitric acid an insoluble residue remains, decomposition procedures must be performed, such as soda potash melts for silicates and barium sulfate, acidic decomposition with potassium hydrogensulfate for oxides of aluminum, iron, titanium and chromium and the Freiberg decomposition with sodium carbonate and sulfur for tin dioxide.

B. Analytical procedures in cation (full) analysis. Each qualitative analysis also includes a quantitative statement which is established by the limiting concentration and detection limits. The limiting concentration (LC) indicates in how many milliliters (of water or other solvent) 1 g of a material can still be detected. As a negative decadic logarithm (analogous to pH), the limiting concentration is called the pD value ($-\log LC$). The detection limit is not a concentration value, but rather an amount. It represents the absolute detectable amount, whereby a solvent drop of approximately 0.05 mL is used as a basis. At a pD value of 6.0, the detection limit is $0.05\ \mu g$.

The classical separation procedure for cations uses group reagents: based on the precipitating agent, the precipitable metal ions are categorized in the hydrogen sulfide group (precipitation from a hydrochloric acid solution), urotropine group (precipitation with ammonia, which is produced during the hydrolysis of hexamethylenetetramine = urotropine; see p. 35), the ammonium sulfide group and the ammonium carbonate group. Alkali metal ions, magnesium ion and ammonium ion remain in the soluble group.

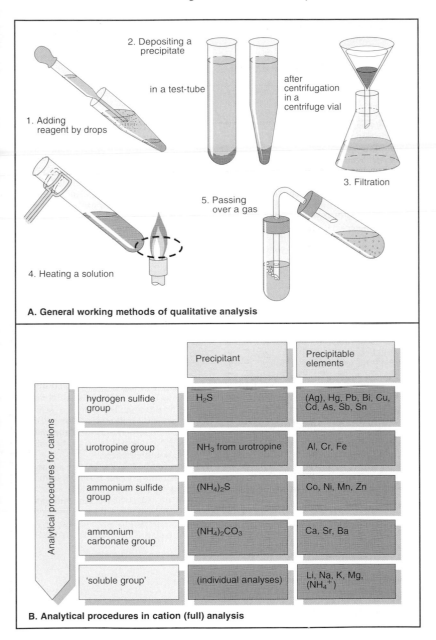

1. Adding reagent by drops

2. Depositing a precipitate

in a test-tube

after centrifugation in a centrifuge vial

3. Filtration

4. Heating a solution

5. Passing over a gas

A. General working methods of qualitative analysis

	Precipitant	Precipitable elements
hydrogen sulfide group	H_2S	(Ag), Hg, Pb, Bi, Cu, Cd, As, Sb, Sn
urotropine group	NH_3 from urotropine	Al, Cr, Fe
ammonium sulfide group	$(NH_4)_2S$	Co, Ni, Mn, Zn
ammonium carbonate group	$(NH_4)_2CO_3$	Ca, Sr, Ba
'soluble group'	(individual analyses)	Li, Na, K, Mg, (NH_4^+)

Analytical procedures for cations

B. Analytical procedures in cation (full) analysis

3.2 Separation Procedures and Selective Reagents

A. Schematic separation procedure for the copper group. In the hydrogen sulfide separation process, whereby the metals Hg, Pb, Bi, Cu, Cd, As, Sb and Sn are jointly precipitated as sulfides from a hydrochloric acid solution, there is a second separation into two groups. As, Sb and Sn, which form soluble thio complexes (e.g. AsS_4^{3-}), are dissolved again using ammonium polysulfide, $(NH_4)_2S_x$, and then they are separated again or detected within the arsenic group (not shown here). The other metals remain as sulfides in the precipitate; they are associated with the copper group. The diagram for the copper group shows the procedure for these classical separation methods, which were largely introduced in courses developed by C. R. Fresenius.

Upon heating with a 20% HNO_3 solution, HgS, S_8 (from the polysulfide by oxidation) and some $PbSO_4$ (SO_4^{2-} by oxidation of S^{2-}) remain; the other metal sulfides go into solution as cations. After the separation of lead as $PbSO_4$, the addition of ammonia until a basic reaction occurs leads to the precipitation of bismuth (Bi) as $Bi(OH)SO_4$, as a basic sulfate. Soluble copper tetraamine (blue) and cadmium tetraamine (colorless) are formed in the process. A selective precipitation of the cadmium besides the copper is accomplished by converting the tetraamine complexes into cyano complexes: copper is reduced, with the formation of dicyan. Owing to the differences in the stability of the complexes, hydrogen sulfide in an ammoniacal solution can be used to precipitate the cadmium from the relatively weak cyano complex as yellow CdS. Copper stays in solution as a Cu(I) cyano complex. In addition to these reactions, additional detection reactions are used for all cations in order to confirm their presence in the separation procedure.

B. Chromate–sulfate procedure for alkali and alkaline earth metal ions. First the alkaline earth metal ions are precipitated together using ammonium carbonate in the course of the cation separation procedure; only magnesium stays in the filtrate along with the alkali metal ions, the 'soluble group.' The carbonates are dissolved in acetic acid; after adding potassium dichromate solution, the yellow barium chromate precipitates from a hot acetate-buffered solution. Carbonate (sodium carbonate) is used again to remove excess chromate ions. The precipitate, from which the chromate is removed by washing, is dissolved in dilute HCl, and the presence of strontium is tested for by using an ammonium sulfate solution. Using ammonium oxalate as a precipitator, calcium cannot be detected in the filtrate of this sulfate precipitation. Thus, the differing solubilities of carbonates, sulfates and oxalates are used for the separation as well as the detection of ions here.

C. Precipitators and metal complexes. Organic precipitators and complexing reagents are suitable for detecting cations and for quantitative analyses (see gravimetry, photometry). As a transition metal, iron combines with ligand atoms such as nitrogen as stable coordination compounds. A ferroin structure (water-soluble red chelate complex) is produced with α,α'-dipyridyl. On the other hand, numerous cations can be precipitated with 8-hydroxyquinoline, depending on the pH. It can be used in the separation stage for detecting magnesium as an oxinate (8-hydroxyquinolinate). Diacetyl dioxime represents a selective nickel reagent: interference due to Fe and Co can be overcome using complexation. The red Ni diacetyl dioxime complex is slightly soluble in water.

Urotropine (hexamethylenetetramine) is a condensation product from ammonia and formaldehyde, which degrades back to its starting material when heated with water. In the presence of NH_4Cl the pH goes to 5–6, which is optimal for precipitating trivalent cations (see p. 33, Section B).

	Hg^{2+}	Pb^{2+}	Bi^{3+}	Cu^{2+}	Cd^{2+}	
$+ H_2S$	HgS	PbS	Bi_2S_3	CuS	CdS	
$+ 20\%\ HNO_3$	HgS	Pb^{2+}	Bi^{3+}	Cu^{2+}	Cd^{2+}	
		$PbSO_4$	Bi^{3+}	Cu^{2+}	Cd^{2+}	$+ H_2SO_4$
			$Bi(OH)SO_4$	$[Cu(NH_3)_4]^{2+}$	$[Cd(NH_3)_4]^{2+}$	$+$ ammonia
				$[Cu(CN)_4]^{3-}$	$[Cd(CN)_4]^{2-}$	$+ CN^-$
				$[Cu(CN)_4]^{3-}$	CdS	$+ H_2S$

A. Schematic separation procedure for the copper group

	Ba^{2+}	Sr^{2+}	Ca^{2+}	Li^+	Na^+	K^+	Mg^{2+}
$+ (NH_4)_2CO_3$ $+ NH_4Cl$	$BaCO_3$	$SrCO_3$	$CaCO_3$	filtrate for 'soluble group'			
$+ CH_3COOH$	Ba^{2+}	Sr^{2+}	Ca^{2+}				
$+ CH_3COONa$ $+ K_2Cr_2O_7$	$BaCrO_4$	Sr^{2+}	Ca^{2+}	$+ Na_2CO_3$			
		$SrCO_3$	$CaCO_3$	$+$ dil. HCl			
		$SrSO_4$	Ca^{2+}	$+ (NH_4)_2SO_4$			
			CaC_2O_4	$+ CH_3COONa$ $+ (NH_4)_2C_2O_4$			

B. Chromate–sulfate procedure for alkali and alkaline earth metal ions

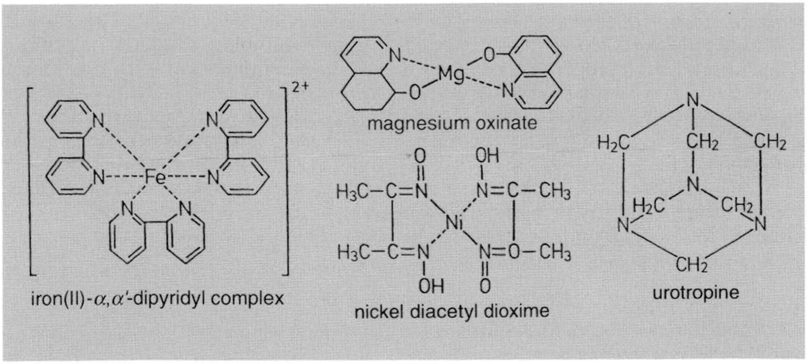

iron(II)-α,α'-dipyridyl complex

magnesium oxinate

nickel diacetyl dioxime

urotropine

C. Precipitators and metal complexes

3.3 Special Test Methods

A. Spot plate (analysis). Spot analysis is a microanalytical procedure. The detection reactions are performed by combining 1 to 2 drops (0.03 to 0.1 mL) each of sample solution and reagent solution in the well-like depressions in a white spot plate made of porcelain. Characteristic colored spots, solutions or precipitates are formed.

B. Test papers for the semi-quantitative analysis of halide ions. Spot analysis can also be performed on spotting paper (cardboard-thick, absorbent, white filter-paper) instead of on spot plates. For example, halide ions can be semi-quantitatively detected by the reactions of a colored silver salt (such as Ag chromate); the paper is impregnated with the Ag compound. White silver chloride is formed. Since a defined drop size from the sample is used, the size of the white circles can be used to estimate the concentration range.

C. Test papers for different metal ions. Here too the test papers are impregnated with selective reagents for the conversion of the metal ions to colored compounds (usually complex compounds). Iron(II) ions react with α,α'-dipyridyl (see p. 35) in mineral acidic solution to form a deep red, very stable complex cation. A deep red precipitate is produced on white test paper by reacting nickel ions with diacetyl dioxime (see p. 35). A largely non-destructive test method results from moistening the surface with a small volume of 'spot acid' (e.g. dilute HNO_3 or HCl with hydrogen peroxide), then pressing the test paper on this location. The limits of detection for non-destructive material testing are about 0.5% (Ni, Co) or between 2 and 25 mg/L in solution (Fe or Co).

D. Test strips with color comparison. Over 300 years ago, the English physicist and chemist Robert Boyle (1627–1691) dabbed or 'spotted' a paper soaked in litmus with a drop of sample solution in order to be able to differentiate acids from bases. The test strips of our day are dipped into the solution to be examined. Square reaction surfaces made of paper on plastic strips contain very small amounts of all necessary chemicals (color reagent, buffer, complexing agent, etc.) for a special selective test reaction: the elements (ions) or dissolved compounds to be detected are detectable owing to color reactions (color changes). The paper soaks up a certain volume of sample solution and therefore a certain amount of material, which then reacts with the reagents present in the paper. The intensity of the coloration, which is evaluated using a color scale, is a measure of the concentration range. The second zone on the nitrate test strip is the nitrite warning zone. To avoid any possible interference by nitrite, a spatula-tip amount of amido-sulfonic acid is added to the sample. The nitrate reaction is based on the formation of an azo dye after the reduction of the nitrate. The nitrite warning zone does not contain any reducing agent.

E. Detector tubes for air analysis. Gas detector tubes are made of glass and are filled with reagents that produce highly selective (specific) reactions or color changes with particular gaseous substances. As a rule the reagents are sorptively bound to a carrier material such as silica gel. A defined volume of air is drawn through the detector tube using a bellows pump. The length of the colored zone corresponds to a particular concentration in the air sample. Detector tubes are calibrated by the manufacturer with gases of defined concentrations. Limiting values such as the 'maximum workplace concentration' (MWC value) are indicated. The number of strokes determines the sample volume and therefore also the detection sensitivity. Since the detector tubes are sealed closed before use, they can be stored for a long time. Several hundred substances in air (or in water and soil after conversion to the gas phase) can be identified and quantified using this analytical technique.

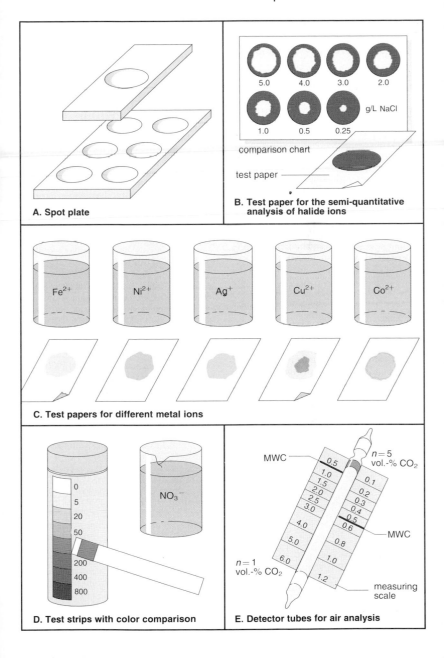

A. Spot plate

B. Test paper for the semi-quantitative analysis of halide ions

comparison chart

test paper

| 5.0 | 4.0 | 3.0 | 2.0 |
| 1.0 | 0.5 | 0.25 | g/L NaCl |

Fe^{2+} Ni^{2+} Ag^+ Cu^{2+} Co^{2+}

C. Test papers for different metal ions

D. Test strips with color comparison

0
5
20
50
200
400
800

NO_3^-

E. Detector tubes for air analysis

MWC

$n=5$ vol.-% CO_2

0.5
1.0
1.5
2.0
2.5
3.0
4.0
5.0
6.0

0.1
0.2
0.3
0.4
0.5
0.6
0.8
1.0
1.2

MWC

$n=1$ vol.-% CO_2

measuring scale

4 Chemical and Biochemical Methods

4.1 Gravimetry

Introduction. If the analytical procedure used to quantify a substance is based on weighings, methodologically it belongs to the field of gravimetry (analysis by weight). As a precipitation analysis method, gravimetry is preferred over other methods for macroanalyses with sample amounts in the gram range, with concentrations greater than 5% (subcomponents or main components) and for selective precipitation methods. Advantages over physical methods include: the absence of calibration, making it an absolute method; a relatively high precision when all influences are taken into consideration; and above all the small outlay on apparatus. A decisive disadvantage for its routine application in operational analysis is the large expenditure of time and effort necessary.

A. Procedures in gravimetric analysis. The analytical procedure consists of sampling (2) from a sample material (1), initial weighing (3), dissolution of the sample (4) and subsequent precipitation according to a prescribed protocol (5), filtering and washing out the precipitate (6) that is produced, drying and, with inorganic precipitates, usually incineration of the filter-paper and annealing for conversion into a defined compound (7) (formula purity). An example is conversion to Fe_2O_3 after precipitation as $Fe(OH)_3 \cdot xH_2O$ and separation of the copper as a tetraamine complex. The procedure concludes with the final weighing of the precipitate and the calculation of the analytical result, e.g. % Fe in an ore sample. The most important tool in gravimetry is the analytical scale, which functions as a damped, single-pan scale with automatic weight support and with a digital display. A macroscale attains an accuracy of 0.1 mg with a maximum load of 200 g, so that an analytical precision of

0.1% can be achieved when weighing more than 100 mg.

B. Filtration, incineration and annealing. After incineration of the filter-paper in a porcelain crucible by annealing up to $1200\,°C$ (Al_2O_3), inorganic precipitates such as $BaSO_4$, $MgNH_4PO_4 \cdot xH_2O$, $Al(OH)_3 \cdot xH_2O$ or $Fe(OH)_3 \cdot xH_2O$ are converted into defined compounds: $BaSO_4$, $Mg_2P_2O_7$, Al_2O_3, Fe_2O_3.
Organic precipitates such as nickel diacetyl dioxime (see Precipitators and metal complexes, Section 3.2.C) can be filtered through glass filter crucibles and weighed in the crucible even after drying to a constant weight. The organic precipitators often display a higher selectivity and a more favorable gravimetric factor than do inorganic precipitators.

C. Impact of solubility. Sources of error in precipitation analysis include especially solubility effects due to isoionic or different ionic additives. Other frequent sources of error in precipitation analysis are supersaturation, colloid formation, aging of the precipitate (usually gravimetrically favorable), post-precipitation (a process which leads to a change in the composition of the precipitate) and the coprecipitation of foreign ions and molecules (such as the precipitator). Particularly favorable are precipitations from a homogeneous solution in which the precipitator is generated by hydrolysis only by heating in the sample solution (e.g. ammonia from urotropine = hexamethylenetetramine, see p. 35).

According to the law of mass action, the solubility of AgCl decreases at first when excess chloride ions are added, but then the formation of slightly soluble chloro complexes is observed (1). The solubility of Ag_2SO_4 changes depending on the kind of ion; with different ionic additions such as magnesium nitrate, it increases greatly with increasing ion strength (2). These effects can be seen during the precipitation itself and when washing out the precipitates.

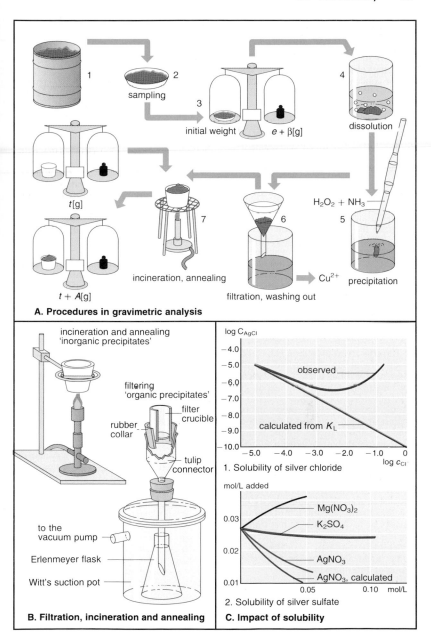

A. Procedures in gravimetric analysis

sampling

3
initial weight   *e* + β[g]

4
dissolution

H_2O_2 + NH_3

t[g]

7
incineration, annealing

6
filtration, washing out

5
precipitation

Cu^{2+}

t + *A*[g]

incineration and annealing
'inorganic precipitates'

filtering
'organic precipitates'

filter
crucible

rubber
collar

tulip
connector

to the
vacuum pump

Erlenmeyer flask

Witt's suction pot

B. Filtration, incineration and annealing

log C_{AgCl}

−4.0
−5.0
−6.0
−7.0
−8.0
−9.0
−10.0

observed

calculated from K_L

−5.0   −4.0   −3.0   −2.0   −1.0   0
log c_{Cl^-}

1. Solubility of silver chloride

mol/L added

0.03

0.02

0.01

$Mg(NO_3)_2$

K_2SO_4

$AgNO_3$

$AgNO_3$, calculated

0.05   0.10   mol/L

2. Solubility of silver sulfate

C. Impact of solubility

4.2 Titrimetric Analysis

Introduction. In titrimetry, materials or groups of materials are quantified by measuring the volume of a reagent solution with known concentrations of a substance, the titrimetric solution which is required for a defined, complete chemical conversion with the materials that are to be analyzed. Adding a reagent until one can recognize (indication) the end-point of the reaction is known as titration. The reagent is called the titrant, and the material to be tested is the sample (or analyte). The chemical conditions that are required as a prerequisite for each titrimetric determination are a defined course of the reaction between the sample and the titrant and the ability to be able to recognize the equivalence point (titration end-point). The type of reaction is based on a systematic scheme within titrimetry: one differentiates between acidimetry (titrations with acids), alkalimetry (titration with bases), redox titrations (transfer of electrons), precipitation titrations (see also gravimetry) and complexometric titrations (complex formation between the sample and the titrant).

A. Burettes. Next to calibrated flasks (e.g. for the production of standard solutions, calibrated for filling at $20\,°C$) and pipettes (for measuring defined volumes as graduated pipettes or transfer pipettes), burettes are the most important piece of basic equipment. They are used for adding the standard solution to the sample solution, either as a single burette with an adjustable discharge (e.g. as a microburette with a volume of $1–5$ mL) or as automatic burettes with automatic zero point setting and a storage vessel for the standard solution. It is possible to read the volume exactly using Schellbach strips: the rear side of the burette consists of frosted glass with a thin blue longitudinal (Schellbach) strip. The point of contact of the mirror images at the meniscus of the liquid appears in the form of two arrows of differing width.

B. Acid–base titrations. Theoretical foundations here include dissociation equilibria, pH values, the ionic product of water, hydrolysis and proteolysis processes, corresponding acid–base pairs and buffering action. Titrimetric analyses are evaluated via a titration curve, which is drawn so as to display the dependence between a value used for indication (pH in this case) and the volume of standard solution added. Acid–base titrations are a good example to show how the course of a titration curve correlates with the sharpness of an indication. Depending on the type of acid–base pair, the jumps in pH occur in the equivalence zone ($t = 1$, degree of conversion) between pH 4 and 10 (strong acid/strong base: I), between pH 7 and 10 (weak acid/strong base: II), at pH 7 (strong acid/weak base: III) or between pH 4 and 7 (weak acid/weak base: IV) The buffering effect of the titration system is obvious with weak acids and bases.

C. Transition points of acid–base indicators. In order to indicate the end-point visually, each acid–base titration requires an indicator which is suitable for this point in the titration curve and which must undergo an easily recognizable color change. Indicators are adapted to the titration system in their mode of action and they themselves act as acids or bases in acid–base titrations. The color indicators which are displayed cover the entire pH range with respect to the transition intervals.

D. Phenolphthalein's mode of action. When a sodium hydroxide solution is added to colorless phenolphthalein with its lactone ring, this ring is opened. Water is split off to form a quinoid ring with a system of conjugated double bonds throughout the entire molecule, whereby the red coloration appears in the pH range $8–10$.

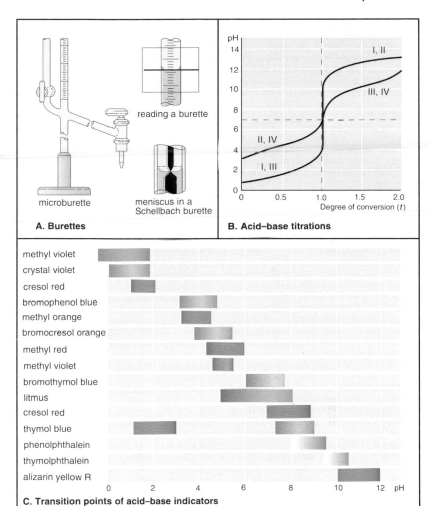

A. Burettes

reading a burette

microburette

meniscus in a
Schellbach burette

B. Acid–base titrations

C. Transition points of acid–base indicators

methyl violet
crystal violet
cresol red
bromophenol blue
methyl orange
bromocresol orange
methyl red
methyl violet
bromothymol blue
litmus
cresol red
thymol blue
phenolphthalein
thymolphthalein
alizarin yellow R

D. Phenolphthalein's mode of action

E. Precipitation titration. Based on the different solubility products of AgI, AgBr and AgCl (8×10^{-17}, 5×10^{-14} and 2×10^{-10}, respectively), theoretically halide ions can be precipitated (fractionated) one after another with silver nitrate. In actual practice, especially with visual indication, usually only one halide can be determined, since with simultaneous titrations the halide that is still soluble is coprecipitated by isomorphic inclusion. With argentimetric precipitation titrations, one can distinguish between two types of visual indication. In the first type, excess titrant forms a colored substance with the indicator [for chloride as per Mohn, silver chromate; as per Volhard, back-titration of excess Ag^+ ions with thiocyanate in the presence of Fe(III) ion and formation of red Fe(III) thiocyanate]. In the second case, adsorption properties of the precipitate are used; the precipitate is positively charged shortly after the equivalence point (isoelectric point) is reached owing to the excess Ag^+ ions. Therefore, negative eosin or fluorescein ions can be adsorptively bound (adsorption indicators as per Fajans). As a result of molecular deformation, intense colorations of the precipitate are generated (eosin is slightly yellow in solution, but is red when adsorbed to AgCl).

F. Redox titrations in general. Oxidation–reduction reactions, or redox reactions, whereby the oxidation number of one or more elements in the sample and titrant change, can also be used titrimetrically by applying the Nernst equation.

The titration curves such as that of an oxidimetric titration (system $Ox_1 + Red_2 = Ox_2 + Red_1$), are generated by calculating the potential E after each addition of the titrant Ox_1 and by plotting this against the percentage conversion. The characteristic points A (starting point), B (50% conversion), C (equivalence point) and D (200% conversion) can also be calculated using the Nernst equation with $E = E^\circ + 0.059/z \log(c_{ox}/c_{red})$ (z = number of electrons exchanged, E° = normal potential). Oxidimetric titrations using permanganate [of oxalic acid, Fe(II), Mn(II), nitrite ions or formic acid, for

example] or using iodine (of arsenite, sulfide, sulfite or H_2O_2) can be indicated directly owing to the titrant's color change (MnO_4^-, red; Mn^{2+}, colorless; I_2 + starch, blue; I^-, colorless). The actual redox indicators are themselves reversible redox pairs whose reduced and oxidized forms are differently colored (e.g. methylene blue, ferroin).

G. Complexometric titrations. These have found very broad application: the tetradentate ligand ethylenediaminetetraacetate (EDTA) forms stable hexacoordinated 1 : 1 complexes with five-membered chelate rings with numerous metals. The molecule is pseudo-octahedrically surrounded by four O and two N atoms, where the N atoms are in *cis*-positions (steric configuration). The stability of the complexes increases with increasing valence of the metal. The complexation equilibrium is strongly pH dependent, since the concentration of free (complexing) ions of EDTA decreases with lower pH values due to protonation. Therefore, a particular pH range must be maintained with direct titration. The more stable the complex, the lower the pH can be. Here also, the indicators are also chelating agents whose constants must be smaller than those of the titrant. Further, the indicator complex has to have a different color to the free indicator. In actual practice, most frequently the sum of Mg and Ca ions (as water hardness) is determined complexometrically. To determine the total hardness, one adds to the water sample an excess of Mg–EDTA solution and then back-titrates the Mg^{2+} ions that are released (mainly by Ca^{2+} ions) and those already present using EDTA. With eriochrome black T as the indicator at pH 10, there is a color change from red (Mg complex) to blue (free indicator).

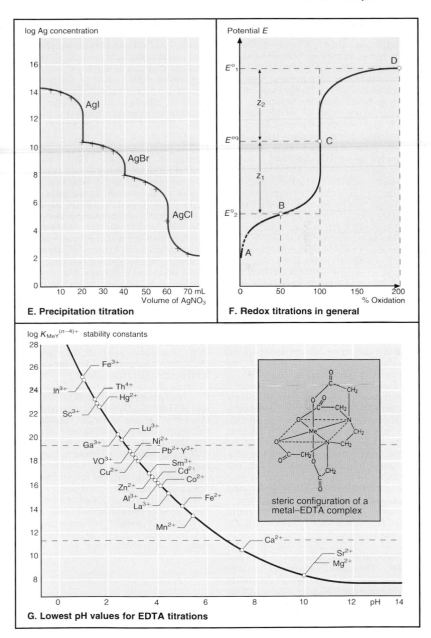

E. Precipitation titration

log Ag concentration — Volume of $AgNO_3$ (mL)

AgI, AgBr, AgCl

F. Redox titrations in general

Potential E — % Oxidation

E^o_1, E^{eq}, E^o_2, z_1, z_2, A, B, C, D

G. Lowest pH values for EDTA titrations

log $K_{MeY}^{(n-4)+}$ stability constants — pH

Fe^{3+}, In^{3+}, Th^{4+}, Hg^{2+}, Sc^{3+}, Lu^{3+}, Ga^{3+}, Ni^{2+}, VO^{3+}, Pb^{2+} Y^{3+}, Cu^{2+}, Sm^{3+}, Zn^{2+}, Cd^{2+}, Al^{3+}, Co^{2+}, La^{3+}, Fe^{2+}, Mn^{2+}, Ca^{2+}, Sr^{2+}, Mg^{2+}

steric configuration of a metal–EDTA complex

4.3 Enzymic Analysis

Introduction. Enzymic analysis determines the concentration of substances using enzymes as well as the catalytic activity of enzymes in biological materials. A unique aspect of analysis with enzymes is the ability to detect individual substances in a complex mixture very selectively (specifically). This obviates the need for time-consuming, labor-intensive and interference-prone separation steps used in composite methods (Section 1.6).

A. Single substrate reaction. Enzymes are biological catalysts, proteins with functional groups where the actual reactions take place. If an enzyme E reacts with a substance, substrate S, an intermediate enzyme–substrate complex is formed. Chemical conversion, such as the transfer of a hydrogen atom, occurs within this complex, which subsequently breaks down to the enzyme and the reaction product P. In many cases, the Michaelis–Menten equation can be used to determine the reaction rate. Prerequisites for the use of this equation are a much higher substrate concentration than enzyme concentration, a rapid first step (large rate constants, k_1) and an irreversible decomposition of the enzyme–substrate complex.

B. Reaction rate of an enzyme–substrate reaction. The reaction rate v is calculated as per Michaelis and Menten, where k is the rate constant of the decomposition reaction. If one determines the dependence of the reaction rate v on the substrate concentration for a single substrate reaction, one arrives at a saturation concentration with v_{max} and the Michaelis constant K_m with $v_{max}/2$. K_m is not an absolute constant; it depends on the pH, the temperature, type of buffer, etc., which must be indicated in the procedural protocol.

C. Structural formulas of NAD and NADP. The most widespread enzyme reactions use NAD- and NADP-dependent dehydrogenases, which, as enzymes of the biological oxidation system, catalyze the transfer of hydrogen. In contrast to the high molecular weight proteins, NAD and NADP are labeled coenzymes (or cosubstrates). As low molecular weight substances, they play a transfer role. From their structure they can be described as pyridine derivatives; they consist of nicotinamide [nicotine = 3-(1-methyl-2-pyrrolidinium)-pyridine], adenine (6-aminopurine) and two ribose moieties linked by a phosphate.

D. Enzymic material determination (end-point method). The end-point method for determining the concentration of a substance can be described using ethanol as an example. In the presence of alcohol dehydrogenase and the coenzyme NAD, ethanol is converted into acetaldehyde, NADH and H^+. Although all enzymic reactions with coenzymes are two-substrate reactions, kinetically they can be treated as single-substrate reactions if one of the substrates is present at a very high concentration relative to its Michaelis constant. Before the enzyme solution is added, the extinction of the sample is measured (photometrically); after the start of the reaction, the change in the extinction is followed until it reaches a constant value; then, after about 3–4 min, the change in extinction is determined and the ethanol level is calculated from this.

E. Absorption spectra of NAD (NADP) and NADH (NADPH). Owing to the light absorption of the pyridine residue, the spectra of NAD and NADH (or NADP and NADPH) display an absorption maximum at 260 nm, and also at 339 nm in a reduced state. Because of these differences, the course of the extinction measurement at 339 nm directly in the photometer cuvette can be used to follow directly the enzymic conversion of the ethanol substrate due to the equivalent amount (or concentration) of NADH which was formed, without interfering with the reaction system.

$$S + E \underset{k_{-1}}{\overset{k_1}{\rightleftharpoons}} ES \overset{k_2}{\longrightarrow} E + P \qquad k_m = \frac{k_{-1} + k_2}{k_1}$$

$$V = K \frac{[E][S]}{[S] + K_m} \text{ with } K_m = (k_{-1} + k_2)/k_1$$

Michaelis-Menten equation

A. Single substrate reaction for substrate (S) and enzyme (E) with enzyme–substrate complex (ES) and Michaelis constant (K_m)

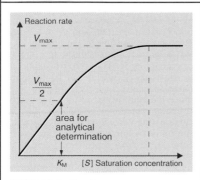

B. Reaction rate of an enzyme–substrate reaction

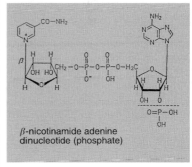

β-nicotinamide adenine dinucleotide (phosphate)

C. Structural formulas of NAD and NADP

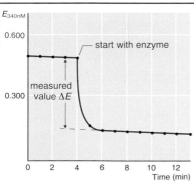

D. Enzymic material determination (end-point method)

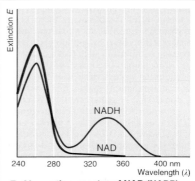

E. Absorption spectra of NAD (NADP) and NADH(NADPH)

F. Reaction scheme for the enzymic analysis of glucose, fructose and mannose. These three sugars are converted by ATP (adenosine triphosphate) in the presence of hexokinase (HK) into glucose-6-phosphate (G-6-P), fructose-6-phosphate (F-6-P) and mannose-6-phosphate (M-6-P) plus ADP (adenosine diphosphate). G-6-P is selectively oxidized to gluconate-6-phosphate by the enzyme glucose-6-phosphate dehydrogenase (G-6-P-DH), whereby the amount of NADPH corresponds to the glucose content. F-6-P and M-6-P are subsequently converted consecutively to G-6-P by phosphoglucose isomerase (PGI) and phosphomannose isomerase (PMI), and they are then converted like glucose. Finally, the consecutively measured extinction values yield the amounts of glucose, fructose and mannose. A pipetting protocol is given for executing this kind of reaction series in enzymic analysis: one after another, 0.10 mL of NADP solution, 0.10 mL of ATP solution and a defined amount of sample solution (e.g. 2.00 mL) are pipetted into a buffer solution (1.00 mL). The solution is mixed, and the extinction is read after 3 min. Then the HK/G-6-P-DH suspension is added and, after mixing again, the extinction is measured again after approximately 10–15 min. Then the PGI and PMI suspensions are added separately, and the extinction values are read.

G. Measurements against the reaction equilibrium. The enzyme reactions represent equilibrium reactions; if there is an unfavorable equilibrium situation, additional reactions must be applied in order to move the equilibrium such that a quantitative conversion becomes possible. A 'regenerating system,' as is used for glutamate determination, makes this possible: the conversion of glutamate with NAD in the presence of glutamate dehydrogenase (GDH) runs to completion at pH 7.6 once the NADH which is formed is continually removed from the equilibrium and is concurrently converted back into NAD using the lactate dehydrogenase (LDH) reaction via pyruvate (salt form of pyruvic acid). As a result of the relatively rapid LDH reaction,

NAD is formed again from NADH in a cycle. The primary reaction ensures that NAD is continually hydrated to form NADH until finally all of the glutamate has been converted. The reaction is stopped by destroying the enzymes, and after the addition of NADH the 2-oxoglutarate which was generated is determined reversibly.

H. Determination of malic acid with secondary reaction. The malic acid salt, L-malate, is oxidized to form oxaloacetate in the presence of L-malate dehydrogenase (L-MDH). To shift the equilibrium to the side of the reaction products, a secondary reaction must be used here as well. This consists of the conversion of the oxaloacetate that is formed with L-glutamate in the presence of glutamate–oxaloacetate transaminase (GOT) to L-aspartate and L-keto(2-oxo)-glutarate. According to the amount of L-malate present, this results in the complete conversion of NAD to NADH (equivalent to the malate), which can be measured photometrically using the spectra in E. Since the equilibrium of the upper partial equation is on the left-hand side, this conversion (the reverse reaction) is used to determine the citrate thus: in the presence of citrate lyase, oxaloacetate and acetate are generated from citrate, oxaloacetate is then reduced to L-malate via NADH in the presence of malate dehydrogenase, whereby NAD is produced again.

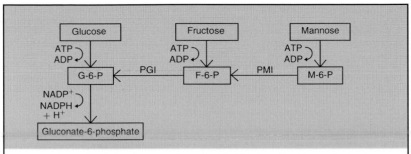

F. Reaction scheme for the enzymic analysis of glucose, fructose and mannose

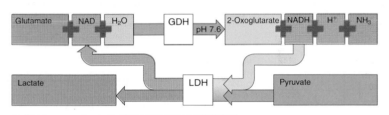

G. Measurements against the reaction equilibrium

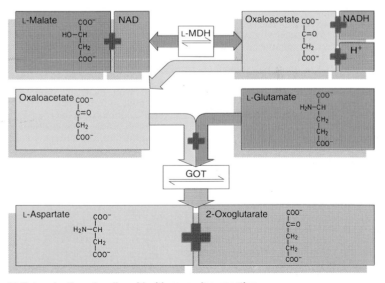

H. Determination of malic acid with secondary reaction

4.4 Immunochemical Methods

Definitions. The term immunoassay includes methods for the determination of biologically active materials, using the specific antigen–antibody reaction. Haptens (low molecular weight substances which are not immunogenic in the body) can be used in place of antigens. Antigens are substances which induce a specific immune reaction in an organism. The interaction between antigen and antibody, or antigen–antibody reaction, consists of a non-covalently linked immune complex, based on the sterically suitable configuration of the antigen to a particular portion (the hypervariable region) of the antibody (lock and key principle).

A. Heterogeneous enzyme immunoassay. The natural ability of the immune system in humans and animals to recognize substances which are foreign to the body and usually have high molecular weights, and to bind these substances specifically with antibodies, can be used analytically. Drugs and hormones have been analyzed in particular using this principle. Owing to progress in antibody technology, it has now become possible to generate specific antibodies against non-immunogenic, or low molecular weight, substances, namely haptens.

In the ELISA (enzyme-linked immunosorbent assay) procedure, antibodies are immobilized on a solid phase, such as the wall of a sample tube (1). After a washing step, an enzyme tracer and the hapten whose concentration is to be determined (2) are added, whereby the hapten concentration increases from left (equal to zero here) to right. The enzyme tracer and the hapten molecules bind competitively to the antibody (3), and an equilibrium (4) is attained. After another washing step (4), which removes the unattached molecules, a substrate is added (5). The substrate only reacts with the enzyme tracer, producing a colored product, the concentration of which (measured by absorption) decreases with increasing hapten concentration (6).

B. Atrazine determination. The procedure just described can be used to determine pesticides, such as the triazine atrazine (1) in this example. The triazines are the haptens. A triazine-carrier protein conjugate consisting of ametryne sulfoxide and hemocyanin (2) is used as the enzyme tracer. The polyclonal antibodies are from rabbits. In the atrazine test procedure, the triazine-specific antibodies are present as a layer (immobilized) on the wall of the lower part of the plastic test vial. If a sample with triazine molecule and immediately thereafter a solution with triazine–enzyme conjugate is introduced into the vial (3), enzyme-labeled triazines compete for binding sites on the antibodies (antibody-atrazine reaction). The solution is removed from the vial, which is washed with distilled water. The colorless solution of a substrate and a chromogen are added (4) in the secondary reaction (enzymic indicator reaction). Now the enzyme–triazine conjugate (and only this conjugate) acts as a catalyst for the conversion of the substrate to a blue pigment. The reaction is particularly sensitive, since a single enzyme conjugate molecule (bound to an antibody molecule) can cause several substrate molecules to react (multiplication principle). The most intense blue coloration is achieved if only enzyme conjugates were present and no triazine molecules (5). However, if one or more substances from the triazine group were in the sample, the blue coloration weakens with increasing concentration. After 2 min, sulfuric acid is pipetted in to stop the color reaction; the color indicator switches to yellow (photometric determination at 450 nm). The calibration is derived from the dependence of the ratio B/B_0 (in % as a ratio of bound tracer in the absence of hapten, B, or the absence of hapten, B_0) on the logarithm of the atrazine concentration (5).

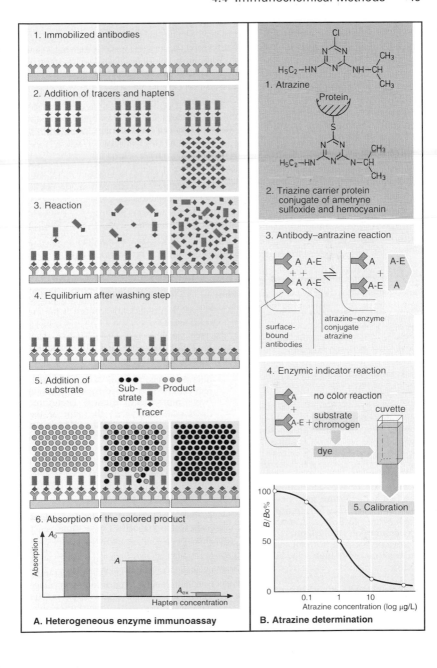

A. Heterogeneous enzyme immunoassay

1. Immobilized antibodies

2. Addition of tracers and haptens

3. Reaction

4. Equilibrium after washing step

5. Addition of substrate

Substrate — Product

Tracer

6. Absorption of the colored product

Absorption — A_0 — A — A_{ex}

Hapten concentration

B. Atrazine determination

1. Atrazine

2. Triazine carrier protein conjugate of ametryne sulfoxide and hemocyanin

3. Antibody–antrazine reaction

A A-E + + A A-E ⇌ A A-E + A-E A

surface-bound antibodies

atrazine–enzyme conjugate atrazine

4. Enzymic indicator reaction

A + A-E + substrate chromogen → no color reaction → dye

cuvette

5. Calibration

$B/B_0\%$

Atrazine concentration (log µg/L)

5 Electrochemical Analytical Methods

5.1 Basic Information

Introduction. Electroanalytical methods use reactions involving ions, electrons and the phase interfaces of electrode and ion conductor to obtain information about the type and amount of inorganic or organic materials. We differentiate between transformations of electric energy into chemical energy (electrolysis with current flow) and transformations of chemical energy into electrical energy (in a galvanic cell). We will use the symbol i for the amperage rather than I, which is usually interpreted as intensity.

A. Electrolysis cell. If one applies a current to a half cell made of a copper rod in a copper ion solution in connection with a second half cell made of a platinum rod in an acidic solution, electrolysis can take place. Copper ions take up electrons and are deposited as the metal on the cathode; the electrons come from the anode reaction.

B. Galvanic cell. A galvanic cell is produced when two half cells, each consisting of a metal electrode and their cationic solutions, are connected to each other. The measurable potential of this cell is the difference between the individual potentials of the electrodes, which are not directly measurable. An oxidation takes place at the anode (zinc ions go into solution), and at the cathode there is a reduction of copper ions to metallic copper, thus generating a current. The two half cells are separated by a diaphragm or frit. The electromotive force (emf) of the galvanic cell can be measured between the poles as the terminal voltage of a galvanic element without current conduction (i.e. with high-impedance resistance).

C. Producing a Galvani potential.

1. Diagram. If one places a metal rod in a solution containing its ions, metal atoms can go into solution as cations by losing electrons (oxidation) and cations can take up electrons from the solution (reduction) and deposit themselves on the rod as metal atoms. An electrochemical or dynamic equilibrium is established in this half cell. The Galvani potential is the non-measurable potential difference between the inside of the electrode and the ions of the adjacent electrolyte solution. The correlation between this potential in the equilibrium and the concentration (or activity) of potential-determining ions is described by the Nernst equation: $E = E^o + (RT/nF) \ln a_{Me}$ (R = universal gas constant; T = absolute temperature; n = charge on the cation; F = Faraday constant).

2. Electric double-layer. A boundary layer model is used to describe the formation of a charged, electric or Helmholtz double layer H (i, inside; o, outside), based on the attractive force between the opposite charges on the metal and in the solution. Excess charges can only pass through this boundary layer to the metal surface as a result of a penetration reaction as a part of the electrode reaction.

D. Potential flow in the electrode boundary layer. The actual electrochemical reaction takes place in the phase interface. The cations which are to be reduced are surrounded by a hydrate shell (water molecules as dipoles). Therefore, in the model one views the sum of these dipoles as a type of charged condenser plate. As a result of the electrochemical process, the immediate environment of the metal loses ions (H_i). a linear potential exists between H_i and H_o as in a condenser; finally a constant potential is attained in the solution—the Nernst potential E.

E. Voltage–current curves for galvanic and electrolysis cells. The terminal voltage as a function of the amperage i amounts to $E^o \pm iR_i$ (R_i = interior resistance), where the sign depends on the type of cells described (see A and B). The decomposition voltage U_z includes the overvoltage η (e.g. as a result of the penetration reaction; see C). It is necessary at least to be able to deposit copper ions electrolytically, for example.

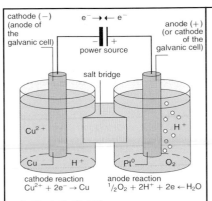

cathode (−)
(anode of
the
galvanic cell)

$e^- \rightarrow \quad \leftarrow e^-$

anode (+)
(or cathode
of the
galvanic cell)

power source

salt bridge

Cu^{2+}

H^+

Cu

H^+

Pt^0

O_2

cathode reaction
$Cu^{2+} + 2e^- \rightarrow Cu$

anode reaction
$^1/_2 O_2 + 2H^+ + 2e \leftarrow H_2O$

A. Electrolysis cell

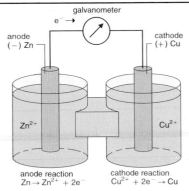

galvanometer

$e^- \rightarrow$

anode
(−) Zn

cathode
(+) Cu

Zn^{2+}

Cu^{2+}

anode reaction
$Zn \rightarrow Zn^{2+} + 2e^-$

cathode reaction
$Cu^{2+} + 2e^- \rightarrow Cu$

B. Galvanic cell

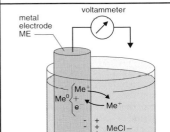

metal
electrode
ME

voltammeter

Me^+

$Me^0 + e$

Me^+

MeCl−
solution

1. Diagram

C. Producing a Galvani potential

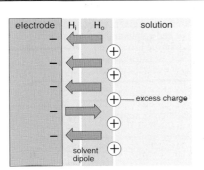

electrode

H_i H_o

solution

excess charge

solvent
dipole

2. Electric double layer

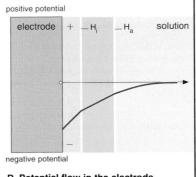

positive potential

electrode

+ H_i H_a solution

negative potential

D. Potential flow in the electrode boundary layer

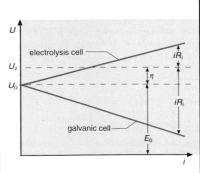

U

electrolysis cell

U_z

iR_i

η

U_0

iR_i

galvanic cell

E_0

i

E. Voltage–current curves for galvanic and electrolysis cells

5.2 Electrogravimetry

Introduction. By using electrogravimetry, materials are quantitatively determined via complete cathodic or anodic deposition (as metals or metal oxides) on electrodes and subsequent weighing. In principle, cathodically there is precipitation of cations with electrons as the precipitant. In contrast to other electrochemical methods, no electrical signal is measured, although the determining step before the weighing is an electrochemical process.

A. Structure of an electrolytic device. In its most general and simplest form, an electrolytic device for electrogravimetry consists of a platinum mesh with a large surface area as the cathode, a platinum coil as the anode (in a rotary form, the anode can also serve as a mixer), a heater, a d.c. power source and a voltammeter and an ammeter as measuring instruments.

There are two modes of electrogravimetry, in the one case the amperage and in the other the potential at the working (platinum) electrode being maintained or directed. If only one electrochemically active and quantitatively separable substance is present in solution, the method with a constant amperage can be chosen. Separations of metals (see C) can only be performed using a reference electrode and a directed controlled potential.

B. Current–voltage diagrams for the electrolysis of a copper sulfate solution. As an example, a 0.1 mol/L copper sulfate solution yields a decomposition voltage (without overvoltage, see p. 51, E) of 0.92 V. To achieve quantitative deposition of the copper (e.g. with a residual concentration of 10^{-6} mol/L), the decomposition voltage U_{ZN} must be increased to 1.07 V according to the Nernst equation (see p. 51, C). However, an overvoltage must also be taken into account, since oxygen gas is produced on the anode (see p. 51, A), and the gas determines the penetration of the hydrogen ions through the boundary layer during the anode reaction. Electrolyses of acidic and basic solutions on platinum electrodes have yielded an almost identical decomposition voltage U_E of about 1.7 V. Thermodynamically (assuming a reversible process), a potential of 1.23 V can be calculated for the dissociation of water, yielding an overvoltage of 0.5 V.

These correlations can be observed in the current–voltage curves: curve 1 shows the theoretical course, curve 2 shows the course taking overvoltages into account, curve 3 described the course with a decreasing copper ion concentration as a result of continuing electrolysis and curve 4 shows the current–voltage curve for hydrogen (on copper). The latter overlays the copper deposits on the platinum and causes the increase seen in the upper portion of curves 2 and 3.

To achieve smaller shifts in the electrolysis potential (from U_E to U'_E) than those shown in the curves, large surface area electrodes are used; the elongation of the depletion layer by mixing is reduced and the diffusion rate increases owing to increased temperatures (see A).

In acidic solutions, the overvoltage of hydrogen depends on the type of electrode and also on the current density. The norm for current density is 5 to 50 mA/cm^2. Although higher current densities provide for more rapid deposition of the metals, they tend to form impure layers with inclusions on the surface of the electrode, and the layers in turn bind poorly to the electrode. In general, the current densities are less than 2 A with common wire gauze electrodes.

C. Decomposition voltage of metal sulfates in 1 mol/L sulfuric acid. As described in Section B, the decomposition voltage changes with the concentration of the metal ions. If the decomposition voltages differ by more than 200 mV, metals can be separated (e.g. Cu and Ag or Ni and Zn) using potentiostatic methods.

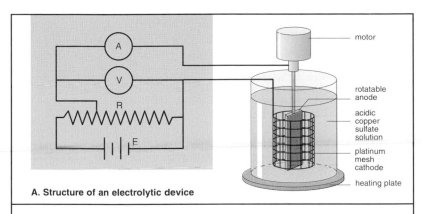

A. Structure of an electrolytic device

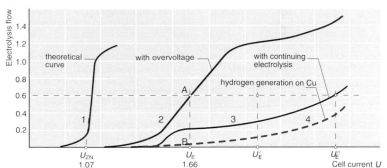

B. Current–voltage diagrams for the electrolysis of a copper sulfate solution

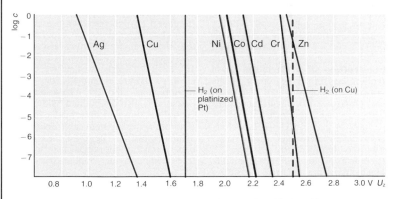

C. Decomposition voltage of metal sulfates in 1 mol/L sulfuric acid

5.3 Potentiometry—pH Measuring Techniques

A. Potentiometric measuring device with wiring diagram. In potentiometry, amounts (or activities) or materials are determined without current by measuring potential differences between a measuring or indicator (working) electrode and a reference electrode in an electrochemical cell. The no-load voltage (cell voltage when no current is flowing) U of the cell represents the sum of the potential differences between the indicator and reference electrodes and of the diffusion potentials. By using a salt bridge, diffusion potentials at the boundaries of the half cells are reduced or largely avoided. When measuring the potential, a voltage is switched counter to the cell voltage so that no current flows. In principle, the voltage can be varied using a linear voltage divider made of high-impedance wire (today this is done using circuits with transistors or microprocessors). The emf (electromotive force) of the cell amounts to $i(R_a + R_g + R_i)$. The galvanometer displays the voltage with R_g. Since R_a and R_g are considerably larger than R_i, R_i can be ignored.

B. Measuring pH with a glass electrode. To measure hydrogen ion concentrations, glass electrodes (as measuring electrodes) are used in combination with a reference electrode such as a saturated Ag/AgCl electrode (see p. 57). The glass electrode consists of a glass tube, usually with a spherical extension on the lower end which contains a buffer solution with a known pH.

C. Construction of a single-rod glass electrode. Glass for pH measurements has an approximate composition of 72% SiO_2, 22% Na_2O and 6% CaO, which is generated by fusing together appropriate amounts of SiO_2, Na_2CO_3 and $CaCO_3$ with a characteristic three-dimensional silicate backbone with end-terminal SiO^- groups.

In practical operation, pH measurements are conducted with single-rod measuring cascades in which the reference electrode is located in the interior of the glass electrode. The glass in the membrane borders with its exterior boundary layer on the measuring solution and with its interior boundary layer with the inner solution with known pH. When the electrode is immersed in a solution, the exterior glass layer swells. Hydrogen ions from the measuring solution penetrate into the swollen layer, whereas sodium ions are displaced owing to their high affinity to end-terminal silicate groups (with their highly negative charge density), similar to an ion-exchange process (see p. 146). Although sodium ions can penetrate out of the sodium silicate skeleton into the gel layer, hydrogen ions cannot advance out of the gel layer to the negative charges of the silicate in the glass. Thus, a potential develops at the glass–water interface.

D. Potential diagram of the glass membrane. The exchange rate between sodium and hydrogen ions on the surface of the glass is pH dependent. Therefore, with different pH values, different potentials also develop inside and out. If there is a lower concentration of hydrogen ions in the measurement solution than in the inner solution (such as 0.1 mol/L), the exterior surface will be negatively charged compared with the interior surface. If the circuit is closed, a charge transport is effected via the sodium ions in the glass—with a kind of impact effect like that with billiard balls—so that the potential within the membrane layer remains constant.

E. Alkaline error of pH glass electrodes. In general, errors occur when measuring pH with glass electrodes if the foreign ions in the measurement solution exceed the hydrogen ion concentration by a factor of 10^{12} to 10^{13}. Then the swollen layer takes up alkali or alkaline earth metal ions such as barium ions, and the phase boundary potential is also determined by their concentration. However, these kinds of alkaline error are not really noticeable until the pH values exceed 10. The size of the error also depends on the ionic radius: metal ions with larger radii introduce smaller errors.

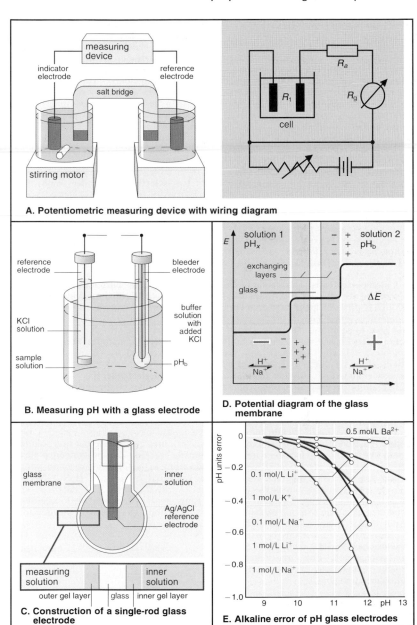

A. Potentiometric measuring device with wiring diagram

B. Measuring pH with a glass electrode

C. Construction of a single-rod glass electrode

D. Potential diagram of the glass membrane

E. Alkaline error of pH glass electrodes

5.3 Potentiometry—Electrodes

Introduction. In potentiometry, the potential difference of an indictor (measuring) electrode is measured against a reference electrode with constant potential in a galvanic cell. According to the Nernst equation (see p. 50), the measured potential difference is a measure of the activity (concentration) of certain ions.

A. Hydrogen electrode. The standard hydrogen electrode is used internationally as the primary reference electrode, whose potential (standard equilibrium Galvani potential) is set equal to 0.000 V. A platinized platinum sheet (coated with amorphous platinum metal to absorb H_2) is present in a solution with hydrogen ions of the measured ion activity of 1 mol/L, with hydrogen gas flowing around it at a pressure of 1.013×10^5 Pa (1 bar).

B. Calomel electrode. This is one of the second kind of electrodes. Electrodes of the first kind consist of a metal submerged in a solution of its ions. Electrodes of the second kind have to equilibria: the metal (Hg) is coated with a thin layer of a slightly soluble compound (Hg_2Cl_2; calomel) and the activity of the Hg_2^{2+} ions is determined by the concentration of chloride (saturated KCl solution) over the solubility product K_L ($2Hg = Hg_2^{2+} + 2e^-$, $Hg_2^{2+} + 2Cl^- = Hg_2Cl_2$; $a_{Hg^{2+}} = K_L/a_{Cl^-}^2$).

C. Ion-selective liquid membrane electrodes. Electrodes which only response to certain free (not bound) measured ions are called ion-selective electrodes (ISE). The earliest ion-selective electrode was the glass electrode (glass membrane electrode) for pH analysis (see p. 54). For liquid membrane electrode with ion exchangers, the potential is determined by the exchange of measured ions (e.g. Ca ions) between the aqueous phase and the fluid, active phase (ion exchanger). The ion-exchanging (organic) counterions are located in the hydrophobic membrane.

D. Redox electrode. Redox electrodes are electrodes of the third kind; they are dependent on the activity of two ions of a redox system from two redox pairs. The best known redox electrode is the quinhydrone electrode, made of a platinum wire submerged in a solution of quinone and hydroquinone. The equilibrium $SbO_4^{3-} + 2H^+ + 2e^- = SbO_3^{3-} + H_2O$ provides the basis for the antimony electrode; because of its pH dependence, it also used to be used as a pH electrode. A third equilibrium (in addition to the two redox pairs) exists between the solution and the solid matter electrode.

E. Gas-selective electrode. With gas-selective (or gas-sensitive) electrodes (gas sensors), gas molecules diffuse through a semipermeable membrane in an outer solution in which an ion-selective electrode is in turn immersed. This makes it possible to determine the CO_2 level by measuring the pH. The outer solution contains a hydrogen carbonate solution, the pH of which varies when CO_2 is introduced, according to the equilibrium between carbon dioxide and hydrogen carbonate ions: $H_2O + CO_2 = HCO_3^- + H^+$. Along with a reference electrode, gas-selective (gas-sensitive) electrodes are single-rod measuring cascades.

F. Enzyme electrode. Enzyme electrodes contain enzymes on the surface of an ion-selective electrode (such as a pH electrode) which has been carefully immobilized in the form of a gel, like in a protein matrix. In the urease electrode (also called a biosensor), urea can diffuse out of a solution into the gel and can be converted there according to the following reaction, which is catalyzed by urease: $H_2NCONH_2 + 4H_2O = 2NH_4^+ + HCO_3^- + 3OH^-$. The change in pH as a result of the formation of hydroxyl ions is a measure of the urea content.

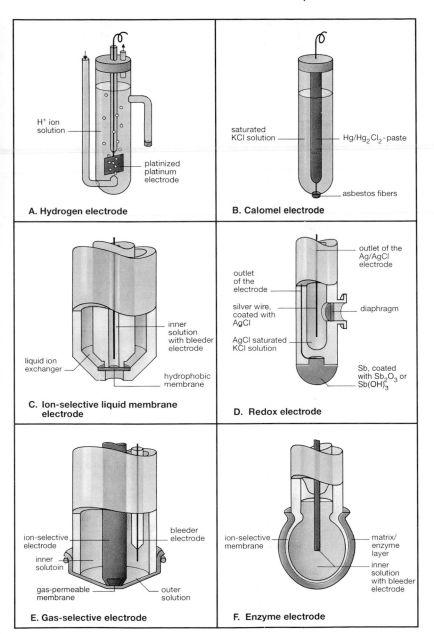

A. Hydrogen electrode

H+ ion solution

platinized platinum electrode

B. Calomel electrode

saturated KCl solution

Hg/Hg$_2$Cl$_2$-paste

asbestos fibers

C. Ion-selective liquid membrane electrode

liquid ion exchanger

inner solution with bleeder electrode

hydrophobic membrane

D. Redox electrode

outlet of the Ag/AgCl electrode

outlet of the electrode

silver wire, coated with AgCl

diaphragm

AgCl saturated KCl solution

Sb, coated with Sb$_2$O$_3$ or Sb(OH)$_3^{-}$

E. Gas-selective electrode

ion-selective electrode

bleeder electrode

inner solution

gas-permeable membrane

outer solution

F. Enzyme electrode

ion-selective membrane

matrix/enzyme layer

inner solution with bleeder electrode

5.3 Potentiometry—Application

Introduction. Direct potentiometry and potentiometric titration are the two areas of application for the use of potential measurements with electrodes.

A. Calibration curve with a calcium-selective electrode. With a liquid-membrane electrode (see p. 56), the potential is measured as a function of free calcium ions. With increasing concentration c, the activity a decreases ($a = fc$; f = activity coefficient, less than 1); only for very dilute solutions does $f = 1$ and therefore $c = a$. Therefore, the calibration curve for the potential measurement as a function of the activity runs below the concentration-dependent curve. According to the Nernst equation (see p. 50), the potential changes by 29.58 mV for each power of ten of the concentration.

B. Calibration curve for a sodium-selective glass membrane electrode. The expanded Nernst equation includes K_{M-I} (M = measuring ion, I = impurity ion) as an empirical constant, which is called the selectivity coefficient. Using a dilution series without any impurity ion, one can derive the detection limit (DL) from the calibration curve; it has an activity $a_I = 0$ for the example of a sodium-selective electrode with approximately 9×10^{-6} mol/L. In the present of impurity ions—here with $a_I = 0.1$ mol/L KCl—the detection limit is reduced to approximately 9×10^{-5} mol/L Na^+. Using the expanded Nernst equation, the selectivity coefficient K_{Na-K} of $9 \times 10^{-5}/0.1 = 9 \times 10^{-4}$ can be calculated (z_M = measuring ion charge, z_I = impurity ion charge).

C. Direct potentiometric fluoride determinations. The fluoride-selective electrode, a solid membrane electrode, has proven to be especially efficient in application. The membrane consists of a thin disk of a lanthanum fluoride (LaF_3) monocrystal in which the fluoride ions conduct the electric current. To reduce the electric resistance, the crystal is doped with europium(II) ions. The fluoride-selective electrode shows a relatively large linear region in the calibration curve (1) which can extend up to fluoride activities of 10^{-6} mol/L. Below this, the solubility of LaF_3 provides noticeable interference.

The fluoride-selective electrode has a high selectivity. However, the pH dependence of the potential measurement, or fluoride pH window, must be taken into account. Both the formation of undissociated HF molecules in the acidic region and interference due to OH^- ions [with partial conversion of the LaF_3 into $La(OH)_3$ and F^-] lead to interference when measuring the potential (2).

D. Potentiometric titration. Simultaneous determinations of halide ions by precipitation with silver ions can be performed using a silver electrode in conjunction with a constant-potential reference electrode. Owing to the differing solubility, the silver iodide or silver bromide precipitates out first; the second equivalence point is attained by AgCl precipitation. At the equivalence point (t = turning point), the solubility product is valid, taking into account any influences on the solubility. The pAg values are the $-\log c(Ag^+)$ values, with V representing the volume of $AgNO_3$ solution with a known concentration that was added. With small solubility differences, a flatter curve is registered (Br^-/Cl^-): the potential jump is considerably less evident than with the I^-/Cl^- precipitation (1). In addition, mixed crystal formation of AgCl and AgBr slides the first equivalence point by more than 2% too far to the right (2). Therefore, in actual practice the simultaneous determination of Cl^-/I^- is better suited to a potentiometric titration. A potentiometric titration is also possible with the complexing agent EDTA (3) with the aid of a calcium-selective electrode.

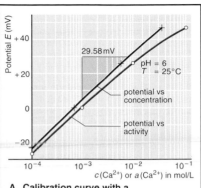

A. Calibration curve with a calcium-selective electrode

29.58 mV

pH = 6
T = 25°C

potential vs concentration

potential vs activity

$c(Ca^{2+})$ or $a(Ca^{2+})$ in mol/L

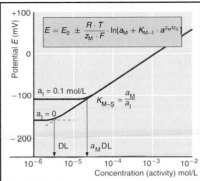

B. Calibration curve for a sodium selective glass membrane electrode

$$E = E_0 \pm \frac{R \cdot T}{z_M \cdot F} \cdot \ln[a_M + K_{M-I} \cdot a^{z_M/z_S}]$$

a_I = 0.1 mol/L

$K_{M-S} = \dfrac{a_M}{a_I}$

a_I = 0

DL a_M DL

Concentration (activity) mol/L

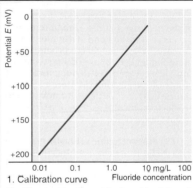

1. Calibration curve

Fluoride concentration

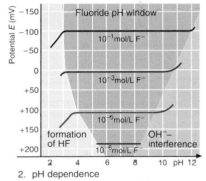

Fluoride pH window

10^{-1} mol/L F$^-$

10^{-3} mol/L F$^-$

10^{-5} mol/L F$^-$

formation of HF

OH$^-$ interference

10^{-6} mol/L F$^-$

2. pH dependence

C. Direct potentiometric fluoride determinations

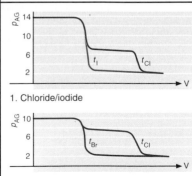

t_I t_{Cl}

1. Chloride/iodide

t_{Br} t_{Cl}

2. Chloride/bromide with AgNO$_3$ solution

D. Potentiometric titration

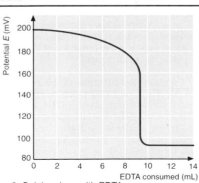

EDTA consumed (mL)

3. Calcium ions with EDTA

5.4 Conductometry

Introduction. In the presence of ions in a solution, current flows if an a.c. voltage is applied. The conductivity G and the resistance R ($R = 1/G$) of this kind of solution consists of the sum of the fractions of the individual ions. Conductometric measurements are mostly useful in analytical applications if in the course of an analytical procedure there are relatively large changes in the conductivity. Therefore, conductometry is principally used as an indication method in titrimetry.

A and B. Principles of measurement and measuring cell. The principle of conductometric analysis is demonstrated using a Wheatstone bridge circuit in which the resistance R_x of a solution is determined. By using a solution having a precisely known conductivity (potassium chloride solution), the ratio of the (plate) distance (l) and plate surface area (q) of the electrodes ($l/q = C$) as the cell constant C can be determined. The upper conductometric measuring cell is suitable for titrations and the lower one is mostly for determining the conductivity of a solution. The specific conductivity can be calculated from the ratio of the cell constant and the resistance ($\kappa = C/R$). To avoid electrochemical processes on the electrodes, the conductivity is always measured using an alternating current. The frequency depends on how large the conductivity is; with electronic measuring equipment, the frequency is automatically reversed with the conductivity range adjustment. Concentrated standard solutions are used in titrations owing to the influence of the fill height on the measurement.

C. Conductivity titration of HCl with NaOH. If a strong acid (such as HCl) is titrated with a strong base (NaOH), a high conductivity will be measured at the start of the titration owing to the hydronium concentration which is present. Up until the equivalence point, there is a displacement of these ions (with the formation of water) due to the less mobile sodium ions, so that the curve drops steeply. After the equivalence point has been passed, the con-

ductivity is determined in an additive fashion from the equivalent conductivities of sodium and hydroxyl ions and of the constant chloride ion concentration, and thus the curve goes back up relatively steeply.

D. Titration of acetic acid (1) or of HCl and acetic acid (2) with NaOH. If a weak acid (e.g. acetic acid) is titrated with a strong base, the sample solution will have a relatively low conductivity because of the slight dissociation of the acetic acid, and at the start the conductivity will decline further owing to the continued decrease of hydronium ion (1). Only in the course of the titration is sufficient strongly dissociated sodium acetate formed that the conductivity from the partial conductivities of the sodium and acetate ions increases. Because of the hydrolysis at the equivalence point, the curve starts to bend in this area. Where the curve is bent, tangents can be laid along the individual branches of the curve, and the equivalence point is derived from their points of intersection. Mixtures of hydrochloric acid and acetic acid can be titrated conductometrically as well due to the different course of the curve (2).

E. Titration of acetic acid with NaOH or with NH₃. If a weak acid (acetic acid) is titrated with a weak base (ammonia), the low initial conductivity increases owing to the formation of ammonium and acetate ions. The hydrolysis of both ions prevents the decrease in conductivity seen with the titration of acetic acid and sodium hydroxide solution. After crossing the equivalence point, the increase is very small because of the slight dissociation of ammonia in water.

Thus, conductometry makes possible titrations similar to those in potentiometry (with pH analyses), but in some cases with curves which can be evaluated better. In addition, precipitation titrations can also be performed conductometrically.

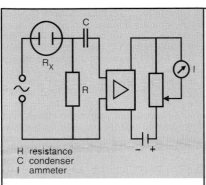

R resistance
C condenser
I ammeter

A. Measuring principle

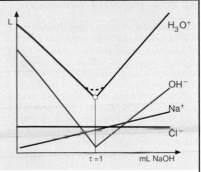

C. Conductivity titration of HCl with NaOH

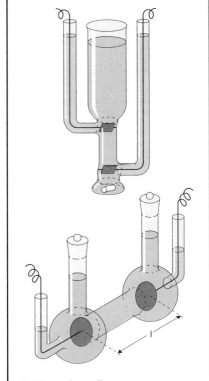

B. Measuring cells

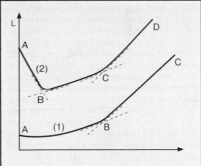

D. Titration of acetic acid (1) or HCl and acetic acid (2) with NaOH

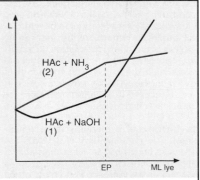

E. Titration of acetic acid with NaOH (1) or with NH₃ (2)

5.5 Polarography and Voltammetry

Terminology. Voltammetry (from 'volt' and 'amperometry') encompasses electrochemical methods with measurement of the current (amperage is indicated as i instead of I in order to avoid confusion with intensity in spectrometric methods) in an electrolytic cell between solid electrodes with a temporally changed current. The current is measured on a polarized reference electrode compared with a non-polarized reference electrode. Polarography is a special type of voltammetry with a liquid (dropping) mercury electrode as the working electrode.

A. Schematic of a simple polarograph with a dropping mercury electrode.

The analytical procedure is based in principle on electrolysis on a microelectrode. Polarography uses a dropping mercury electrode (DME). In its simplest form it consists of a glass capillary which is attached to the mercury storage vessel by PVC tubing. The height of the vessel, and therefore the dropping rate, can be regulated. The analytical equipment consists of a direct current source, a potentiometer (with a synchronous motor to change the current in a time-dependent manner), a galvanometer and the polarographic analysis cell with dropping mercury and reference electrode (in the simplest case; an Hg layer as a deposited layer for the vessel; otherwise usually with a calomel electrode). Before performing polarographic or voltammetric analysis, care must be taken to provide for the complete removal of the interfering oxygen in the analytical solution; at a half-wave potential of 0 or -1 V, it is reduced to water via hydrogen peroxide by two-electron reduction. Also, by adding a conducting electrolyte one must provide for sufficient conductivity.

B. Lifetime of a mercury drop (snapshot).

The glass capillary of a DME 12–15 cm in length has an inner diameter of approximately 0.05 mm. Hg drops grow on the end of the capillary as a result of the mercury pressure. Drop hammers are used for a rapid technique and differential pulse polarography (see p. 66). Good drop properties are needed in polarography (enhancement is possible by siliconizing the capillary). The time-dependent drop size is described by the capillary constant $K = m^{2/3} t^{1/6}$ ($m = $ mass of Hg/s in mg/s; $t = $ dropping time; the lifetime of a drop is between 1 and 3 s).

C. Voltammetric current–voltage diagrams.

The characteristic common to both polarography and voltammetry is the changeable current on two electrodes as the excitation signal. The type and amount of depolarizers (electrochemically active materials to be analyzed) are determined from the course of the current–voltage curve. An increase in current is registered when electrons are transferred at the boundary layer between electrode and solution (Helmholtz or electrochemical double layer): the process is called the penetration reaction. Background current flows at any potential even without a penetration reaction at the reference electrode. If all electrochemically active substances are reacted near the electrode, the diffusion process determines the current flow; the diffusion current in the current–voltage curve is reached. The type of the substance is characterized by the half-wave potential. Step-like current–voltage curves are obtained in polarography. The positions of the half-wave potential and of the diffusion current are obtained by drawing the tangents (background and diffusion current) and the inflectional tangents (AB). The diffusion current is proportional to the concentration of the depolarizer. The shape of the curve can be traced back to the special process of surface growth and movement of the particles to this surface. Voltammetry on resting (stationary) electrodes leads to pointed current–voltage curves with a peak current and peak potential (2).

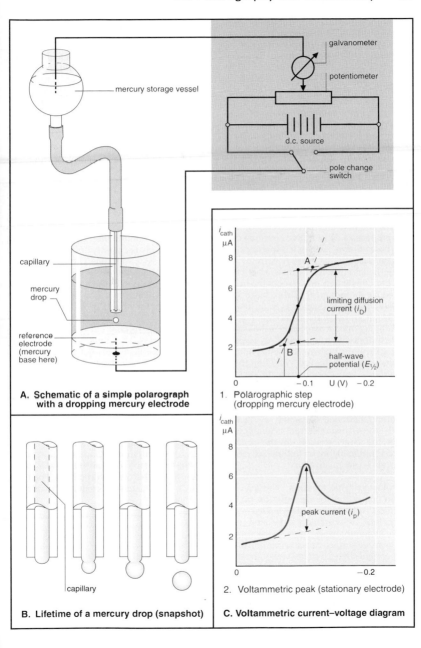

A. Schematic of a simple polarograph with a dropping mercury electrode

1. Polarographic step (dropping mercury electrode)

limiting diffusion current (i_D)

half-wave potential ($E_{1/2}$)

B. Lifetime of a mercury drop (snapshot)

2. Voltammetric peak (stationary electrode)

peak current (i_p)

C. Voltammetric current–voltage diagram

D. Current–voltage curves determined polarographically. In classical d.c. polarography, a current–voltage curve is recorded as a polarogram with a dropping mercury electrode. The advantage of the dropping electrode lies in the continually recurring, reproducible renewal of the Hg drop, i.e. of the electrode surface. Owing to the surface growth of the electrode (see p. 63) and the movement of the particles to this surface, a step-like current–voltage curve is produced. Also, the growth of the Hg drop leads to convective flow processes in the immediate environment around the drop. The phase boundary between the metal and electrolytic conductor can be viewed as a condenser (Helmholtz double layer), whereby the double-layer capacity changes with the potential and the electrode surface. This produces a capacitative charging current (base current), which is superimposed on the test signal, the Faraday current. The double-layer capacity is also associated with the convective flow processes in the immediate vicinity of the drop: more ions reach the electrode than by pure diffusion. As a result, polarographic maxima ('without damping') occur especially in the area of inclined slope above the step. They make it more difficult to evaluate the signal height. Damping can be achieved by adding to the electrolyte ('with damping') in the form of high molecular weight substances such as gelatins or surface-active substances. The increasing base current can be traced back to the aforementioned capacitative current. Finally, the diffusion current is attained, at which the concentration of the reacted material immediately on the surface of the electrode is practically zero.

E. Polarographic maxima. The maxima due to convection of the solution on the electrode surface are also called drop oscillations or maxima of the first type (curve 2): inhomogeneities in the current on the electrode surface cause variations in the electric field strength at the base of the drop due to screening and therefore differing interfacial tension of the mercury against the electrolyte. The surface of the electrode is set into motion and the adhering electrolyte layer is entrained. The added substances which are adsorbed on the surface of the electrode have a compensating effect on the inhomogeneities in the field and they prevent the formation of maxima (curve 2). Maxima of the second type (non-interrupting maxima) can be traced back to a too strong discharge of the mercury from the capillary (curve 3). They arise on the diffusion current and extend over a large potential range.

F. Polarogram of a three-substance mixture. Polarography is also suitable for determining several ions (A, B, C) side by side, as long as the half-wave potentials are sufficiently different from one another. An example of this is the simultaneous determination of Tl(I), Cd(II) and Ni(II) in an electrolyte consisting of 1 mol/L each of ammonia and ammonium chloride. Cu, Pb and Cd can also be jointly polarographed in 0.1 mol/L KCl solution. The selection of the electrolyte, the base solution, determines the position of the half-wave potential and therefore the possibility of simultaneous analysis.

G. Current–voltage curves with different electrodes. The course of current–voltage curves is decisively determined by the selection of the electrode. In 1 mol/L KCl solution, for a 10^{-4} mol/L Tl^+ ion concentration, the polarograms or voltammograms are shown for a gold electrode coated with Hg (curve 1), a rotating platinum electrode (as a wire with a thin constant diffusion layer, curve 2) and a dropping mercury electrode (curve 3). After a drop falls, the oscillation maximum is attained, i.e. the base of the curve, at the end of a 'drop life,' shortly before the drop falls. The first two curves can best be evaluated quantitatively.

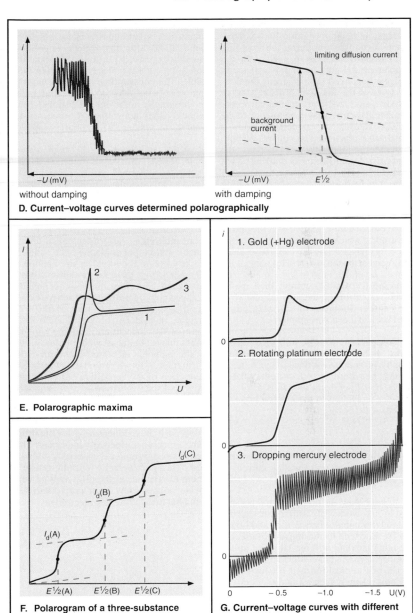

D. Current–voltage curves determined polarographically

without damping

with damping

limiting diffusion current

background current

E. Polarographic maxima

F. Polarogram of a three-substance mixture

G. Current–voltage curves with different electrodes

1. Gold (+Hg) electrode

2. Rotating platinum electrode

3. Dropping mercury electrode

H. Comparison of Faraday and capacity voltage. In the base solution for a polarographic determination, during the recording of a current–voltage curve, even without a depolarizer, a slight current flow will occur, which is generated from charging the double layer on the electrode. This capacitive current i_C (or capacitive charging current or non-Faraday current) is in opposition to the Faraday current i_F associated with a penetration reaction. The courses of both currents run opposite one another: the capacitive current decreases until the end of the life of an Hg drop, whereas the Faraday current increases until it reaches a limiting value, the limiting diffusion current. These regularities are utilized in tast and pulse polarography, in which the current is measured only at a time interval δt at the end of the drop, in which the surface growth of the Hg drop and therefore the capacitive charging current are the lowest. This also yields a more favorable ratio of the Faraday and charging (capacitive) current than in the measurement of the average current over the entire lifetime of an Hg drop (= classical d.c. polarography).

I. Comparison of pulse and differential pulse polarography. In pulse polarography, in the course of the voltage increase during the recording of a current–voltage curve shortly before the drop falls, a voltage impulse is applied to the electrochemical cell:

$$\frac{dE_{dc}}{dt}$$

at a time in which the difference between the capacitive current and Faraday current is the greatest.

1. Voltage flow. In normal pulse polarography, one works only with square signals. There is an impulse between 30 and 60 ms per Hg drop with increasing amplitude. On the other hand, in differential pulse polarography a small square-wave voltage of 50 to 100 mV at most for 5 to 100 ms superimposes a slowly increasing direct current voltage toward the end of the Hg drop's lifetime. The lifetime of the Hg drop is between 2 and 3 s.

2. Current–time curves. With both techniques, the current intensity is not registered until 20 ms after the voltage impulse is switched on in order to allow the capacitive charging current to decay as much as possible. The current–time curves for normal pulse polarography or differential pulse polarography differ only in region a–b, in which either no voltage (normal) or an increasing voltage (differential) is applied.

3. Pulse polarograms. In both cases, polarograms with small single steps are generated. Normal pulse polarography shows polarograms similar to those in classical d.c. polarography. However, the sensitivity of the latter technique is 10-fold greater. The polarograms are often better formed than in classical polarography.

In differential (also often called difference) pulse polarography, the current shortly before the impulses are applied is deducted from the total current measured afterward. Therefore, one does not obtain a stepped polarogram here, but rather a Gauss (bell)-shaped curve. Instead of the step height, the peak height is used as the measure of the concentration of the analyte. With this technique as well, the sensitivity of the method is increased as a result of the formation of the differential and a further reduction of the capacitive current intensity, and the evaluation is also improved owing to the formation of maxima.

When performing these special polarographic techniques, a synchronous pulse superposition and measuring for the dropping process are especially important. This requirement is attained especially with the aid of tast polarography (current measurement at intervals at the drop end) or with pulse polarography with drop-controlled (and not free-falling) Hg electrodes.

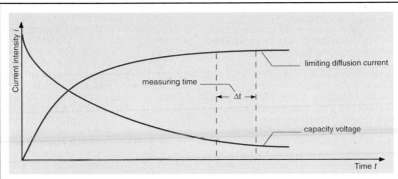

H. Comparison of Faraday and capacity voltage

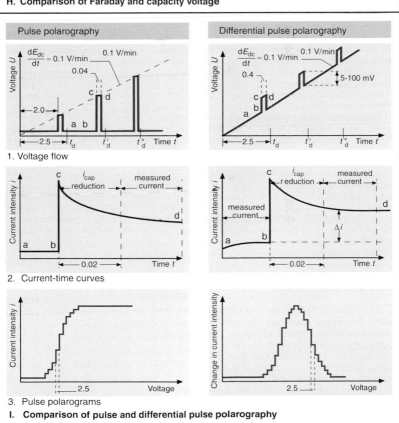

Pulse polarography

1. Voltage flow

2. Current-time curves

3. Pulse polarograms

Differential pulse polarography

I. Comparison of pulse and differential pulse polarography

J. Tast a.c. polarography of a Cd^{2+}/Zn^{2+} ion solution. In contrast to d.c. polarography, in a.c. polarography a small (5–50 mV) and low-frequency (5–100 Hz) alternating voltage superimposes the usual d.c. polarographic measuring circuit. If the Helmholtz double layer is still viewed as a condenser, then compared with the Faraday a.c., the capacitive a.c. is phase shifted. Therefore, in a phase-selective measurement, the capacitive current can be largely suppressed. The detection limits of a.c. polarography are lower than those of d.c. polarography. For example, by using tast a.c. polarography and with a dropping mercury electrode (DME), analyses of Zn and Cd can be conducted side by side until far below 10^{-4} mol/L. Measurements of nA currents (instead of μA as in the figure) are technically possible without any difficulties. Because of the good resolution of the signals (with a difference of 500 mV in the peak potentials), both metals can be measured side by side even with extremely differing concentration ratios.

K. Differential pulse polarography of maleic and fumaric acid. Numerous organic substances having electrochemically active (reducible or oxidizable) functional groups can be determined polarographically just as sensitively and selectively as inorganic ions. The example shows that even *cis–trans* isomers such as maleic and fumaric acid can be determined separately using differential pulse polarography. With signal heights of several hundred nA, concentrations in the lower milligram range per liter are accessible. The bottom curve shows the current–voltage curve of the electrolyte (base) solution which was used.

L. The principle behind inverse voltammetry (stripping method). In inverse techniques, an enrichment of the electrochemically active substance on the electrode precedes the actual measurement. For a metal cation, voltammetry (on a stationary electrode) leads cathodically to a particular peak current (direct determination: upper curve). The height of the peak depends on the concentration of the metal in an amal-

gam, for example, with a stationary Hg electrode. One can attain a higher sensitivity, i.e. considerably higher signals, if the separation of the metal in the Hg drop (amalgam formation) is performed for a particular time at a constant electrolysis voltage. This pre-electrolysis is performed with a potential which is approximately 200 mV more negative than the half-wave potential or peak potential. Afterwards, the working electrode potential is varied in the anodic direction with a constant rate of change, bringing the metal back into solution in the process. This results in a considerably larger inverse signal. Because of this subsequent redissolution, inverse working methods are also known as stripping techniques.

M. Inverse differential pulse polarography of cations in drinking water. Especially low detection limits can be achieved using a combination of differential pulse polarography with the inverse technique described above. This method is also called DPASV, or differential pulse anodic stripping voltammetry. This involves the trace determination of picogram amounts (10^{-12} g) of Cd, Pb and Cu, and of nanogram amounts of zinc. Following a pre-electrolysis of 2 min at -1.3 V (against an Ag/AgCl electrode), curve 4 shows the blank value of the basic electrolyte from an ammonium citrate buffer (pH 4.7); curve 3 shows the sample in the basic electrolyte, and curves 1 and 2 are each for the sample after the first or second addition of a standard solution (standard addition method, see p. 16).

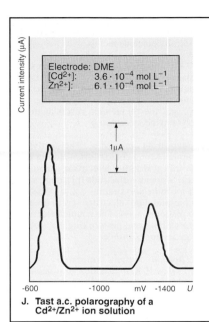

J. Tast a.c. polarography of a Cd²⁺/Zn²⁺ ion solution

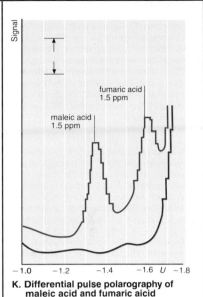

K. Differential pulse polarography of maleic acid and fumaric aicid

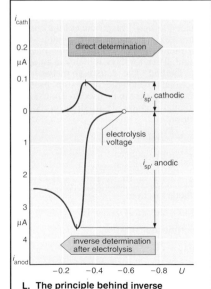

L. The principle behind inverse voltammetry (stripping method)

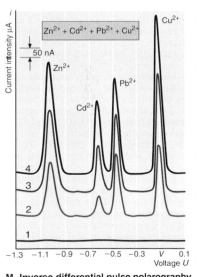

M. Inverse differential pulse polarography of cations in drinking water

5.6 Coulometry

Introduction. Compared with electrolysis and electrogravimetry (see p. 52), in coulometry the electrons are not merely used in reactions, but rather their quantity is measured for a special reaction corresponding to Faraday's law. The material conversion is proportional to the quantity of electricity. The current efficiency must be 100% for the method to be used.

A. Potentiostatic coulometry. In potential-monitored coulometry, the potential of the working electrode is monitored in comparison with a reference electrode, and it is kept constant. This potential is slightly higher than the corresponding half-wave potential; thus this method has tie-ins with electrogravimetry and with polarography.

1. Measuring cell with three-electrode arrangement. The coulometric cell consists of a working electrode, made of platinum here, a reference electrode and a counter-electrode. The potentiostat ensures constancy of the potential of the working electrode compared with that of the reference electrode. The counter-electrode is separated from the working electrode by a diaphragm. The electrolysis cell must be constructed such that the analyte can be converted 100% on the working electrode.

2. Current–time curves. To evaluate current–time curves, the e function is used to show the current intensity as a function of time. The amount of charge used can also be determined from the surface integral. The actual measuring instrument used in this method is an electronic integrator. To be able to follow the entire electrolysis, one must electrolyze up to a current intensity which is less than 0.1% of the initial intensity $i_{t=0}$. Then more than 99.9% of the material is deposited. To reduce the long analytical time required, the integrator is used to register the current intensity only at certain times, and the natural logarithm of the intensities is plotted as a function of time. In this way one obtains a line with a slope having a constant k and initial current intensity $i_{t=0}$ as the intercept with the ordinate. This method is useful for determining small amounts of non-iron metals, especially precious metals.

B. Galvanostatic coulometry. In galvanostatic coulometry, or coulometric titration, the titrant is generated electrochemically at a constant current in the first step, and the titrant then reacts with the analyte in the second step.

1. Measuring apparatus. The analytical equipment for a coulometric titration consists of a working electrode (platinum sheet), a counter-electrode (platinum wire) and an indicator system, e.g. a polarized double-platinum electrode consisting of two platinum wires. The counter-electrode is the cathode and the working electrode is the generator electrode.

2. Titration curve: reversible system $2Br^-/Br_2$. If KBr is added to a solution, bromide is deposited on the working electrode, or anode, as a reagent for conversions. As long as bromide is consumed, the polarization potential U is measured.

At the end-point is the reversible system $2Br^- = Br_2$ and the polarization potential is reduced to zero. The product of the current intensity i and the time t of the jump to the end point yields the volume of material sought. Therefore, coulometry is an absolute method. The detection method just described is called the 'dead-stop' process. This procedure can be used to titrate small amounts of ammonia in slightly alkaline solution:

Anode reaction:

$$2Br^- = Br_2 + 2e^-$$

$$Br_2 + 2OH^- = Br^- + BrO^- + H_2O$$

Cathode reaction:

$$2H_2O + 2e^- = 2OH^- + H_2$$

Determination reaction:

$$2NH_3 + 3BrO^- = 3Br^- + N_2 + 3H_2O$$

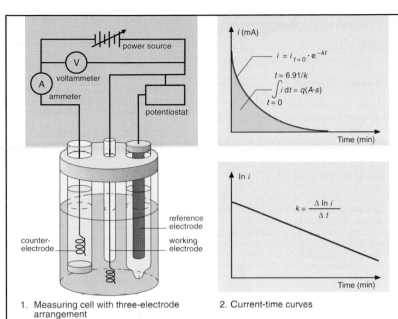

$$i = i_{t=0} \cdot e^{-kt}$$

$$t = 6.91/k$$

$$\int_{t=0} i\,dt = q(A \cdot s)$$

$$k = \frac{\Delta \ln i}{\Delta t}$$

1. Measuring cell with three-electrode arrangement

2. Current-time curves

A. Potentiostatic coulometry

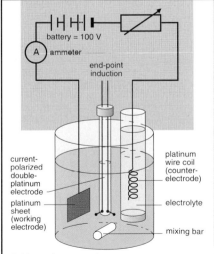

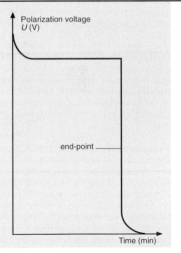

1. Measuring apparatus

2. Titration curve: reversible system $2\,Br^-/Br_2$

B. Galvanostatic coulometry

6 Thermal Analytical Methods

6.1 Methods–Overview

Definition. Thermal analysis (TA) is a general term for methods in which physical or chemical properties of a substance, a mixture and/or reaction mixtures are measured as a function of time or temperature, whereby the sample is subjected to a controlled temperature program (DIN 51005–1983).

A. Thermogravimetry (TG). In thermogravimetry, the change in a sample's mass is measured in the course of a preset temperature–time program. A change in mass occurs when volatile substances are formed in the thermal reaction. These would include water (as water vapor), carbon dioxide and similar substances. In addition to decompositions, oxidations can also occur, depending on the atmosphere. The measurements are made using a thermal balance. Each step in the thermogram corresponds to a particular reaction and can be assigned to the appearance of a particular substance.

B. Derivative thermogravimetry (DTG). Instead of the step-like curve seen in thermogravimetry (A), in derivative thermogravimetry one sees the first derivative of curve A, shown here for a two-stage reaction in which water and carbon dioxide are released.

C. Enthalpimetry (calorimetry). Calorimeters are used to measure heat. For historical reasons, the measurement of enthalpies of combustion processes under an oxygen overpressure in calorimetric bombs is often still referred to as calorimetry instead of enthalpimetry. The curve shows the results of an experiment in an adiabatic calorimeter: the temperature difference δT is a measure of a sample's physical calorific value.

D. Differential thermal analysis (DTA). In this method the temperature difference between an analyte and a known reference sample is examined. Both samples run through a preset temperature–time program. Positive signals indicate a conversion with the release of energy (heat); negative signals show that energy (heat) is being used: exo- and endothermic reactions.

E. Differential scanning calorimetry (DSC). Differential scanning calorimetry is a method in which the difference of the addition of energy to a substance and a reference material is measured as a function of temperature while the substance and reference material are subjected to a regulated temperature program. In contrast to differential thermal analysis, the measuring signal is recorded proportional to the difference of the output.

F. Thermometric titration. Chemical reactions often include the uptake or release of energy: when acids react with bases, the heat of neutralization is released. In thermometric titration, measurements of the change in temperature are used as an indication of titrimetric determinations. The temperature (change) is registered as a function of the titrator volume.

G. Emission gas thermoanalysis (EGT). Emission gas thermoanalysis is a thermoanalytical method in which the type and/or amount of volatile gaseous products are determined. For detection, the thermal conductivity measurement can be used as an additional analytical principle based on thermal processes.

H. Thermomechanical analysis (TMA). A sample is subjected to a temperature program in which it is subjected to a constant or changing force. Changes in either volume or dimension are measured.

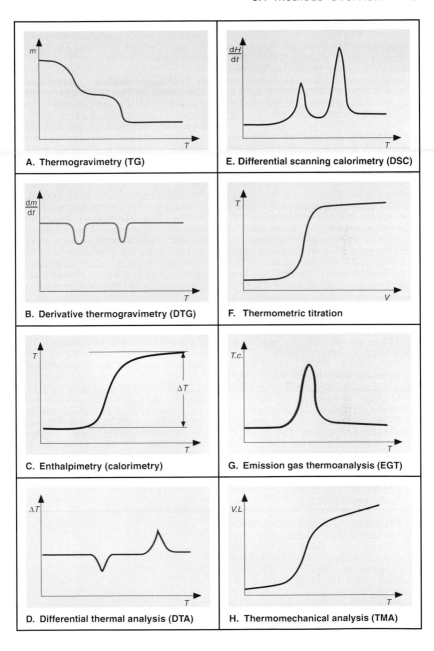

A. Thermogravimetry (TG)

B. Derivative thermogravimetry (DTG)

C. Enthalpimetry (calorimetry)

D. Differential thermal analysis (DTA)

E. Differential scanning calorimetry (DSC)

F. Thermometric titration

G. Emission gas thermoanalysis (EGT)

H. Thermomechanical analysis (TMA)

6.2 Thermogravimetry

Introduction. Mass changes recorded in thermogravimetric analysis (TGA) can be explained based on physical or chemical processes: as a result of decomposition, oxidation, dehydration or drying and surface reactions, whereby the surrounding atmosphere must be considered.

A. Schematic diagram of a thermogravimetric analysis device. To conduct a TGA, the analyte is heated in a crucible (sample container) using an oven following a prescribed program and in a particular atmosphere which can be regulated using a gas inlet and outlet. The sample container is connected to a balance which consists of the suspension, a weighing beam made of quartz and a counterweight. Automatic zero point balances have an electronic sensor which detects deviations of the balance beam from the zero position using a lamp with a window and a photocell. Compensation can be done electromagnetically; a corresponding change in the current intensity is proportional to the change in mass and is recorded. Temperature measurements can be taken directly next to the crucible using a thermoelement.

B. Release of water from $CaC_2O_4 \cdot H_2O$. The packing of the sample and the geometry of the sample container have a significant effect on a TGA curve. At a heating rate of $10\,°C/min$, the release of water from calcium oxalate monohydrate ($CaC_2O_4 \cdot H_2O$) was recorded using a mass spectrometer. Arrangements 1 and 2 produce symmetric curves as a result of uniform hydration. The highly asymmetric curve 3 can be traced back to delayed and inhibited dehydration due to the type of packing.

C. Classification of thermogravimetric curves. The physical and chemical changes which took place under the prescribed temperature changes can be detected from the shape of the TGA curve. Drying to eliminate the water totally can lead to a stable product. Stepped curves show the presence of two or three stable forms. An irregular shape indicates a thermally unstable substance. Reactions with the surrounding environment are noticeable at the start of a TGA curve, and can lead to a stable product.

D. Thermogravimetric simultaneous determination. One of the areas of application for TGA is in the expansion of conventional gravimetry (see p. 38) to determine simultaneously materials which were coprecipitated, such as the oxalates of calcium, strontium and barium, without prior separation. The TGA curve starts with the total mass of the dried precipitates $MeC_2O_4 \cdot H_2O$ (1), at $250\,°C$ the water of crystallization (2) of all three oxalates is liberated, at $400\,°C$ there is a joint decomposition (3) in which carbon monoxide is released and carbonates (4) are formed. The two following signals can be separately assigned to the decomposition of calcium carbonate (4) or strontium carbonate (5) (carbon dioxide release). Barium can be determined as the difference from the total mass. The upper curve represents the first derivative of the TGA curve (derivative thermogravimetry).

E. Characterization of polymers. TGA plays an important role in the characterization of polymers: the decomposition or transition starts at different temperatures and shows a different curve for poly(vinyl chloride) (PVC) and the aromatic polypyromellitimide (PI). Poly(methylene methacrylate) (PMMA), high-pressure polyethylene (HPPE) and polytetrafluoroethylene (PTFE) differ in their decomposition temperatures.

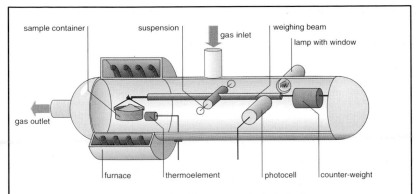

A. Schematic diagram of a thermogravimetric analysis device

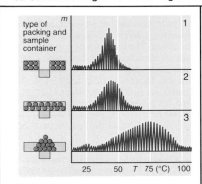

B. Release of water from $CaC_2O_4 \cdot H_2O$

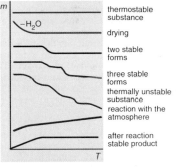

C. Classification of thermogravimetric curves

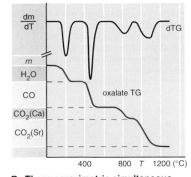

D. Thermogravimetric simultaneous determination

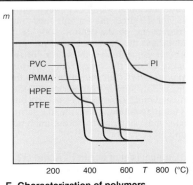

E. Characterization of polymers

6.3 Differential Thermal Analysis (DTA)

Principle. Differential thermal analysis measures temperature differences while a sample is heated or cooled in comparison with a reference substance which shows no thermal effect in the temperature range being examined. Endothermic and exothermic conversions and chemical reactions are measured with this method (see also pp. 74–75).

A. Make-up of a DTA measuring cell. The apparatus for DTA consists of a measuring device for the temperature differences to be determined, a thermal oven with a temperature control unit, amplifier and recorder and a unit for monitoring and controlling the atmosphere in the oven. Primarily thermoelements are used for measuring temperatures between -170 and $2500\,°C$ of the sample, the difference between the sample and the reference material $T_S - T_R$ and of the oven. The arrangement of the temperature-measuring elements influences the shape of DTA curves.

B. Sample mount. The shape of the DTA curve is also influenced by the sample mount. In the first case, the sample and inert reference material are located in different chambers of a metal block; in the second case they are arranged next to each other on ceramic rods. The metal block has symmetric boreholes. Metal or ceramic crucibles for holding the sample and inert substance are on the ceramic rods. If the sample and the inert substance are positioned in separate crucibles and not in a heating block, the equalization of temperature occurs via convection and is thus considerably slower than in a metal block. This analytical equipment is sensitive, but it has poor resolution. Therefore, the sample mount must be adapted to the analytical problem.

C. Comparison of normal TA and DTA. In differential thermal analysis (DTA) there is no slope to the curve, and positive and negative signals are easier to evaluate (2). In general, endothermic effects are induced by a phase change, dehydration, reduction and some kinds of decomposition reactions. The causes of exothermic effects are largely crystallization, oxidation and also decomposition reactions.

At the start of an exothermic reaction, the evolution of heat from the sample increases (1) and the thermal flow from the sample's environment decreases. When both thermal flows are equal in size, the maximum of the signal in the DTA curve has been reached. The inert (reference) substance has a considerable influence on the analytical result: it should be as close as possible to the analyte in its thermal properties (thermal conductivity). On the other hand, in the sample's reaction range the inert substance should not show any thermal reaction. Aluminum oxide and magnesium oxide are suitable as reference substances in many situations. The limits for an optimal heating rate are determined by the reaction rate in the sample and by the thermal conductivity (by the formation of a temperature gradient). As in TGA, consideration should also be given to the composition of the atmosphere, the gas pressure and the initial weight (decisive for the heat capacity of the measuring system).

D. Characteristic signal forms of DTA. A physical and a chemical transition differ in the width (sharpness) of their signal. Also, a decomposition of the sample can be recognized by a zig-zag-shaped curve. The signal area depends on the mass of the reactive sample, on geometric factors of the sample, its thermal conductivity and the reaction enthalpy. The width and shape of a DTA signal are influenced by the type of transition and by a number of mechanical factors (see above).

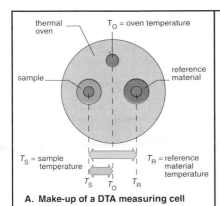

thermal oven

T_O = oven temperature

sample

reference material

T_S = sample temperature

T_R = reference material temperature

T_S T_O T_R

A. Make-up of a DTA measuring cell

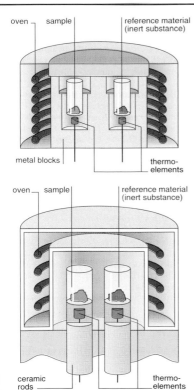

oven   sample   reference material (inert substance)

metal blocks   thermo-elements

oven   sample   reference material (inert substance)

ceramic rods   thermo-elements

B. Sample mount

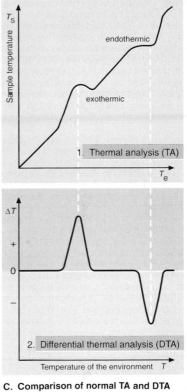

T_S

Sample temperature

endothermic

exothermic

1. Thermal analysis (TA)

T_e

ΔT

+

0

−

2. Differential thermal analysis (DTA)

Temperature of the environment T

C. Comparison of normal TA and DTA

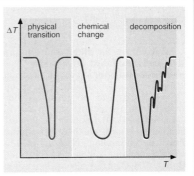

ΔT

physical transition   chemical change   decomposition

T

D. Characteristic signal forms of DTA

E. DTA microanalysis of bladder and urinary stones. The analytical applications of DTA (and DSC) extend from the determination of phase diagrams (especially for clay minerals, salt minerals and metal alloys) for comparative studies and identifying materials to the examination of ignition temperatures and ignition processes in explosives, purity testing for inorganic and organic substances and the characterization of polymers, natural products and even biological materials. With DTA curves it is possible to differentiate between bladder stones and urinary stones based on phosphate, oxalate or urate stones; this differentiation is important for specific therapy. It is also possible to differentiate between calcium and magnesium phosphate stones without chemical analysis in favorable situations.

F. DTA curves for seven polymers in a mixture. In the schematic course of a DTA curve (a) with transition points for a polymeric organic substance, one can observe the glass transition (1), a crystallization (2), the fusion process (3), an oxidation process (5) [dashed line (4) is without oxidation] and the decomposition (6). With DTA, signals can be unambiguously assigned in a mixture of seven polymers (b) (melting points in parentheses):

1 polytetrafluoroethylene (20 °C),
2 high-pressure polyethylene (108 °C),
3 low-pressure polyethylene (127 °C),
4 polypropylene (165 °C),
5 polyoxymethylene (174 °C),
6 nylon 6 (220 °C) and
7 nylon 66 (257 °C).

G. DTA curves for alkali and alkaline earth metal sulfates. DTA yields characteristic and differing curves for alkali and alkaline earth metal sulfates: e.g. gypsum ($CaSO_4 \cdot 2H_2O$) can be readily differentiated from the double salt glauberite [$K_2Ca(SO_4)_2 \cdot H_2O$]. Likewise, a clear differentiation is possible between sodium sulfate as mirabilite ($Na_2SO_4 \cdot 10H_2O$) and magnesium sulfate as epsomite ($MgSO_4 \cdot 7H_2O$). DTA also makes it possible to detect a different number of molecules in the water of crystallization, for example between epsomite and kieserite with only one H_2O molecule. The most differentiated DTA curves are seen for the complex double salts of sodium magnesium sulfates: of $Na_2Mg(SO_4)_2 \cdot 4H_2O$ and $Na_{12}Mg_7(SO_4)_{13} \cdot 15H_2O$ (loeweite).

H. DTA curves for natural clay minerals. Qualitative differences in natural clays can be determined using DTA and a qualitative and also to a limited degree a quantitative phase analysis. Metamorphosis-based differences in the structure of clay minerals can be very clearly recognized from DTA curves. DTA enhances X-ray and electron microscopic studies. Similarities are seen for kaolin and antigorite, both of which represent two-layer clay minerals with octahedral Si_2O_5 groups. Montmorillonite and talc are three-layer clay minerals. Illite belongs to the micaceous group and has a totally different DTA behavior with barely recognizable signals.

I. Differentiation of microorganisms. Various microorganisms can be differentiated from DTA curves owing to their organic contents (metabolic products). Samples of 20 mg are analyzed with 20–40 mg of calcined kaolinite in an oxygen atmosphere (with a flow rate of 10 mL/min). The characteristic maxima are at 280/445/532 °C for *Corynebacterium diphtheriae*, 272/292/427/546 °C for *Streptomyces* spp. and 264/309/465/652 °C for *Escherichia coli*.

It is also possible to differentiate between *Aerobacter aerogenes* cultured in C- and N-limiting media, likewise via the shape and the differing maxima of the DTA curves at 280/310/481 and 255/322/452 °C, respectively.

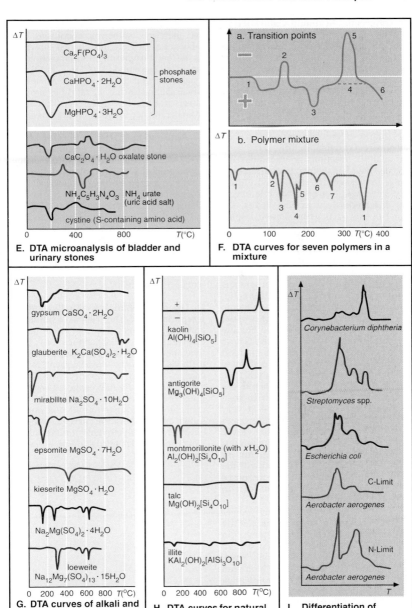

E. DTA microanalysis of bladder and urinary stones

F. DTA curves for seven polymers in a mixture

G. DTA curves of alkali and alkaline earth metal sulfates

H. DTA curves for natural clay minerals

I. Differentiation of microorganisms

6.4 Differential Scanning Calorimetry

Principle. In differential scanning calorimetry (DSC, also called dynamic differential calorimetry, DDC), the sample and a reference substance are kept at the same temperature by adding heat, and the heating rate (dH/dT) required to do so is measured as a function of temperature or time. This method is especially suitable for quantitative studies of the released or absorbed enthalpy of a reaction. In differential thermal analysis, reactions or transitions are recorded based on temperature differences (see p. 76) between the sample and a reference substance. However, these temperature differences lead to heat fluxes, the course of which has a considerable impact on the DTA curve.

A. Differential scanning calorimeter. Calorimeters are used to measure heat. The type of operation wherein a sample can be subjected to a temperature–time program is called dynamic or scanning. A differential scanning calorimeter (also called a dynamic heat flux differential calorimeter) consists of a 'twin calorimeter' with two heating elements having a low heat capacity. These heating elements heat the sample and inert substance (reference material) according to a uniform, pre-determined program. The scanning aspect of the method lies in the comparisons of the temperatures in both measuring cells, which are done at very short intervals. Temperature differences lead to a change in the heating power via the control system and therefore to a continuous temperature equalization. Aluminum or other metal vessels in direct contact with the heating element and the temperature sensor are used as containers for amounts of 1 to 100 mg of substance.

B. DSC measuring curves. The electric signals based on different heating powers in endothermic or exothermic processes are recorded and cited in joules per unit time. With exothermic processes which run quickly, the programmed heating rate must be greater than the temperature change of a sample. Similar problems can occur with rapid endothermic reactions if maximum heating does not insure a linear heating rate and, therefore, isothermic conditions. The influence of the gas atmosphere is shown by the different isothermic decompositions of nickel tetracarbonyl [$Ni(CO)_4 \cdot 2H_2O$] in a dynamic output-difference calorimeter, in which the measuring signal ΔT is proportional to the heat flow difference (heat flows from the furnace to the samples) (curve 2).

C. Application of DSC to the characterization of deoxyribonucleosides. Examples are described in the literature of the application of DSC in biological systems for the characterization of biopolymers for biochemical and clinical-chemical studies. Signal maxima between 400 and 600 K indicate the different deoxyribonucleosides 2′-deoxyuridine, thymidine, 2′-deoxycytidine, 2′-deoxycytidine · HCl, 2′-deoxyadenosine · H_2O and 2′-deoxyguanosine · H_2O. After prior isolation, they can be identified with the help of differential scanning calorimetry.

D. Quantitative analysis of polyethylene terephthalate. The great similarity between DSC and DTA is a reason for the comparatively similar applications. The fundamental difference between the two methods lies in the measurement of the changes in thermal capacity with DSC, which is especially interesting as a typical glass transition in polymers (1). The applicability in quantitative analysis is shown by the calibration curve (2) using an example of polyethylene terephthalate, which fuses under nitrogen with an endothermic reaction.

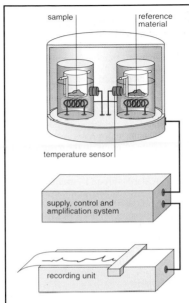

A. Differential scanning calorimeter

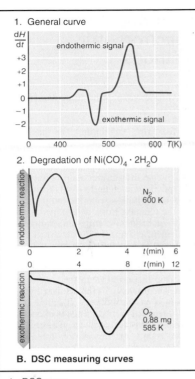

1. General curve

2. Degradation of $Ni(CO)_4 \cdot 2H_2O$

B. DSC measuring curves

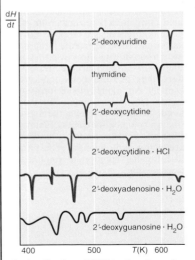

C. Application of DSC to the characterization of deoxyribonucleosides

2'-deoxyuridine

thymidine

2'-deoxycytidine

2'-deoxycytidine · HCl

2'-deoxyadenosine · H₂O

2'-deoxyguanosine · H₂O

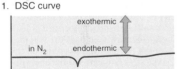

1. DSC curve

exothermic

in N₂   endothermic

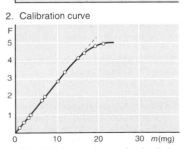

2. Calibration curve

D. Quantitative analyses of polyethylene terephthalate

7 Atomic Spectrometric Methods

7.1 Atomic Absorption Spectrometry (AAS)

A. Measuring principle. The basis of atomic absorption spectrometry is the resonance absorption in gases: if a polychromatic beam of light is transmitted through a gas in an atomic state, radiation of very specific wavelengths from this light is absorbed. A hollow-cathode lamp (HCL) with the element to be determined as the cathode generates the emission spectrum of the element as a result of an electrical glow discharge. The analytical sample, which is in a gaseous state and the atoms of which absorb in the region of the resonance line, is located in the atomizer in the path of the beam. The intensity of the primary light is weakened in the process. Spectral lines which are not absorbed are not weakened. A monochromator is used to examine the region round the resonance line. The weakening of the resonance line is recorded by the detector and finally it is printed out in the form of the inverse spectrum. The intensity of the resonance absorption is directly correlated with the number of particles N to be absorbed according to the Lambert–Beer law (see p. 106). The intensity I_0 of the primary light is reduced thus: $I = I_0 \exp(-kel)$, where k is the absorption coefficient. The flame area (see B) can serve as the absorption cell; it does not have any limited spatial expansion like the glass cuvettes in spectrophotometry (see p. 108).

B. Chemical-physical processes in flame technology. The most common atomizing tool in AAS (next to the graphite furnaces shown in C) is the flame, e.g. the air–acetylene flame, which can reach temperatures up to approximately 2550 K. In contrast to emission analysis (see p. 94), in this case the flame has the task of vaporizing and thermally decomposing the sample. One sprays the solution containing the analyte in special nebulizers with mixing chambers (cf. flame photometry) and conducts the resultant fine spray of particles into the burner using the gas flow. Larger droplets are deposited as a condensate. The fuel gas (e.g. acetylene) and an additional oxidant (e.g. air) are mixed in the burner. In the flame, the solvent water vaporizes as a gas and solid particles in the aerosol remain in the gas flow. Oxides can undergo an undesirable reaction. The target for AAS analysis lies in the generation of atoms in the gas phase. A reduction in sensitivity and spectral disturbances (see also O) occur when molecules (in their ground state), excited molecules, ions, excited atoms and radicals form. The object of optimizing the flame is to reduce these secondary reactions to a minimum. Thus, the goal of atomization is to bring the highest possible portion of the element to be determined into an atomic state, including under reducing conditions if necessary (e.g. in an acetylene–dinitrogen oxide flame).

C. Diagram of an AAS device with graphite furnace. An AAS device consists of: a hollow-cathode lamp for producing the emission spectrum; an apparatus for background correction (here deuterium background correction; see P and Q); the atomizer unit (here a graphite furnace; see G) with a borehole to allow insertion of the sample, with the necessary connections to a power supply (as a heat source); and an inert gas supply. An 'atomic cloud' forms in the atomizer. The monochromator has the task of single-handedly shielding the region of the main resonance line (see A). Finally, the absorption signal is recorded by the detector as a reduction in the intensity of the primary light following appropriate amplification and conversion.

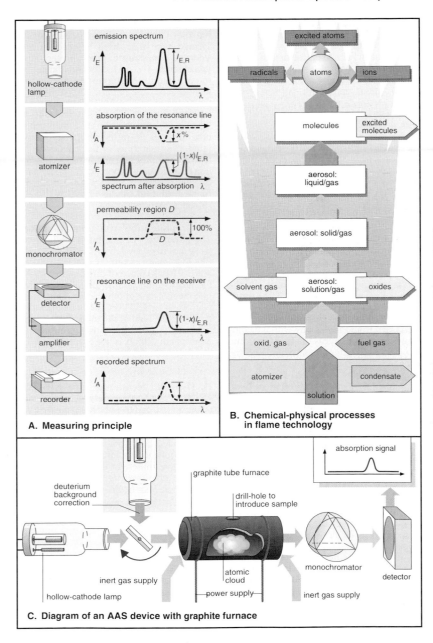

A. Measuring principle

emission spectrum

absorption of the resonance line

spectrum after absorption

permeability region D

resonance line on the receiver

recorded spectrum

hollow-cathode lamp

atomizer

monochromator

detector

amplifier

recorder

B. Chemical-physical processes in flame technology

excited atoms

radicals   atoms   ions

molecules   excited molecules

aerosol: liquid/gas

aerosol: solid/gas

solvent gas   aerosol: solution/gas   oxides

oxid. gas   fuel gas

atomizer   condensate

solution

C. Diagram of an AAS device with graphite furnace

absorption signal

deuterium background correction

graphite tube furnace

drill-hole to introduce sample

monochromator

detector

hollow-cathode lamp

inert gas supply

power supply

inert gas supply

atomic cloud

D. Hollow-cathode lamp (HCL). The high selectivity of AAS is due to the fact that a special hollow-cathode lamp is used for each element. Each lamp consists of a glass cylinder filled with a noble gas (neon or argon) at a pressure of a few mbar, with a fused cathode and anode (with a quartz window through which UV rays can pass). The cathode, in the shape of a half cylinder with a glass container (for protection), is made of the metal which is to be analyzed (or is filled with the metal). The anode is a thick wire made of tungsten or nickel. When several hundred volts are applied, there is a glow discharge which leads to the generation of the emission spectrum of the particular element.

E. Mixing chamber burner. It is necessary to have the longest possible light path (1) through the absorbing flame zone in order to increase the sensitivity. This is made possible by using slit burner heads 10 cm long. The light path runs in the longitudinal axis of the flame. The sample solution is pneumatically atomized into a mixing chamber. The aerosol which is created is led from there, along with the combustible gas mixture of fuel gas and oxodizing gas after intermixing, into the expanded flame of the slit burner. An acetylene–air mixture with rates of combustion of around 1.6 m/s and temperatures around 2000 °C (approximately 2300 K) works especially well. Higher temperatures are attained with acetylene–oxygen or acetylene–dinitrogen oxide flames (around 3000 °C). Larger droplets are deposited on surfaces of impingement in the mixing chamber; only the fine aerosol reaches the flame.

F. Construction of a graphite tube furnace. A graphite tube furnace can be used to atomize the sample instead of flames: as per Massmann, a 28 mm long graphite tube (inner diameter 6 mm) is held by two cooled graphite contacts, through which the current is applied. Temperatures of 3000 K can be reached by resistance heating at 8 V and 400 A. The argon inert atmosphere prevents the graphite tube from burning. The light emitted by the HCL shines through the quartz window. The sample solution is fed into the tube from above using a microliter syringe, then it is dried based on a preselected temperature program (see also processes in C), and is heated further to remove interfering impurities in a thermal manner before the actual atomization. After the atomization (see M) the tube is baked before the next analysis. The advantages of flameless (electrothermal) AAS in graphite tube furnaces lie in the ability to remove interfering materials in the course of the temperature program and in the concentration of the atomic vapor. Although the path of light is shorter than in a flame, in the furnace dilution of the vapor due to the solvent is avoided.

G. Graphite tube with L'vov platform. The advantages of graphite tube furnaces were described by L'vov as early as 1958 (the Massmann cuvette was not developed until 1970). In place of direct placement in the furnace, the sample can also be dosed into the tube in a graphite boat. This L'vov platform is heated principally by the radiation from the wall of the tube; the atomization takes place after a time delay in an atmosphere which is thermically already stabilized. A cut-away section is shown here.

H. Hydride technique. The elements antimony, arsenic, bismuth, selenium, tellurium and tin can be volatilized as hydrides from a solution after reduction with $NaBH_4$ and they can thereby be freed from interfering materials. The atomization takes place in a heated quartz cuvette. Mercury can be determined as the sole element in metallic form (because of the high vapor pressure of 0.0016 hPa at 20 °C).

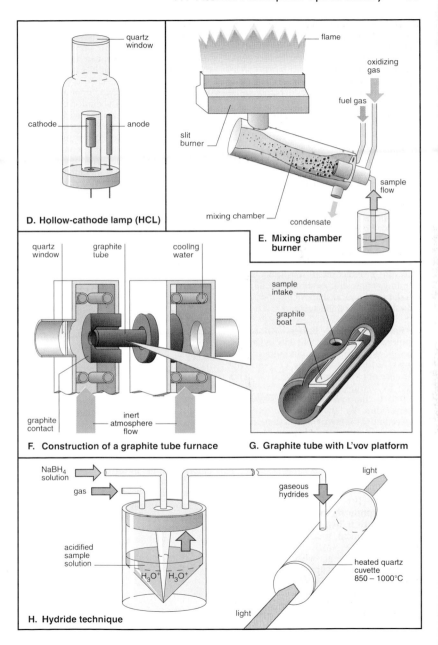

D. Hollow-cathode lamp (HCL)

quartz window

cathode   anode

E. Mixing chamber burner

flame

oxidizing gas

fuel gas

slit burner

sample flow

mixing chamber   condensate

F. Construction of a graphite tube furnace

quartz window   graphite tube   cooling water

graphite contact

inert atmosphere flow

G. Graphite tube with L'vov platform

sample intake

graphite boat

H. Hydride technique

NaBH₄ solution

gas

acidified sample solution

H_3O^+ H_3O^+

gaseous hydrides

light

heated quartz cuvette 850 – 1000°C

light

I. Influence of the spectral bandwidth.
The usable spectral region of AAS lies
between about 190 and 900 nm (193.7 nm
for As; 852.1 nm for Cs), which is the same
wavelength region as used in UV/VIS
spectrophotometry. Relative to the resolution and dispersion, greater demands must
be placed on the monochromators (see B).
The spectral bandwidth plays an especially
important role, for example.

1. Emission spectrum of Si HCL. In AAS
the monochromator has the task of separating the resonance line of the element to be
analyzed from other emission lines from the
radiation source (the example here is silicon
with a wavelength of 251.61 nm). In general, this task can be achieved with a slit
width of 0.2 nm.

2. Absorption signals in the flame. With a
reduction of the spectral slit width below
0.07 nm, a reduction in sensitivity and a
bending of the calibration curve occur. Here
we show the typical shape of the absorption
signals when the flame technique is used.

J. Spectrochemical buffer. At higher
temperatures, elements with low ionization
energy such as rubidium and other alkali
metals occur mostly as ions. Owing to the
presence of other easily ionizable elements
such as sodium, the equilibrium can be
shifted to the side of the undissociated
atoms due to the electron pressure
(according to the law of mass action). In this
case the solution of a sodium salt is also
called a 'spectrochemical buffer.' The high
sodium levels do not interfere, owing to the
method's selectivity: (1) without added Na
and addition of (2) 500, (3) 1000 and (4)
2500 mg/L of Na.

**K. Absorption signals when a graphite
tube furnace is heated.** We differentiate
between chemical, spectral and ionization
interferences in AAS. Chemical interference occurs, e.g., when stable compounds are formed with components of the
flame gas or the sample (example: the formation of phosphate when determining
calcium; this can be suppressed by a concurrent reaction of $LaPO_4$ formation with

the addition of lanthanum). Ionization
interference can also be suppressed (see J).
Spectral interference in the form of an
absorbing background can occur in the
flame and with the graphite tube furnace
technique ('flameless' AAS), e.g. from high
salt concentrations. Two important causes,
the generation of a dissociation continuum
using NaCl as an example and in the dissociation of $MgSO_4$, and the generation of
scattered radiation are shown in the three
figures. In addition, in the gas phase (especially in the flame as well; see C) compounds such as oxides or radicals can be
formed, which leads to broad absorption
bands. The techniques for measuring or
eliminating these non-specific background
absorptions are outlined in M–T (see pp.
88–91).

**L. Temperature program for graphite
tube furnace AAS**
1. General procedure. The individual steps
involved in flameless AAS consist of drying
the sample, ashing and atomization (see
also C). Drying is done at temperatures
around 100 °C; a signal is recorded in the
AAS device due to the light scattering on
water droplets. The cause of the signal
during incineration (temperature near
1200 °C) can lie in the dissociation continuum of SO_3 (see K), for example. The
actual signal of interest occurs during atomization (temperatures between approximately 1800 and 2600 °C, depending on the
element).

2. Real sample. The example of cadmium
determination in urine shows that the
background signal interferes with the
atomic signal with a predefined temperature
program T (extract from drying stage up to
atomization at 1800 °C). It is an important
task of the AAS method to eliminate these
background interferences.

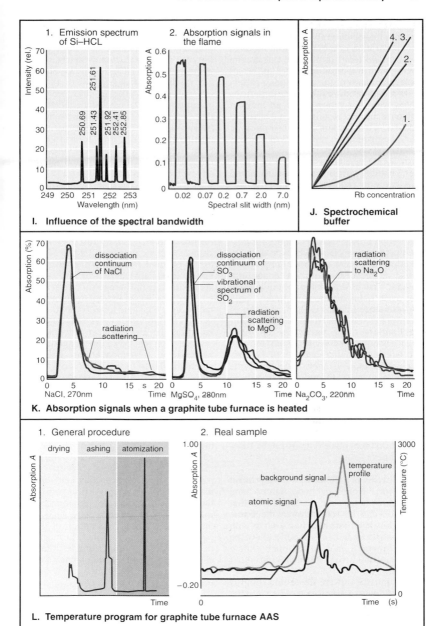

1. Emission spectrum of Si–HCL

Intensity (rel.)

250.69 251.43 251.61 251.92 252.41 252.85

Wavelength (nm)

2. Absorption signals in the flame

Absorption A

Spectral slit width (nm)

I. Influence of the spectral bandwidth

Absorption A

4. 3.
2.
1.

Rb concentration

J. Spectrochemical buffer

Absorption (%)

dissociation continuum of NaCl

radiation scattering

NaCl, 270nm   Time

dissociation continuum of SO_3
vibrational spectrum of SO_2
radiation scattering to MgO

MgSO₄, 280nm   Time

radiation scattering to Na_2O

Na₂CO₃, 220nm   Time

K. Absorption signals when a graphite tube furnace is heated

1. General procedure

Absorption A

drying   ashing   atomization

Time

2. Real sample

Absorption A   Temperature (°C)

temperature profile
background signal
atomic signal

Time   (s)

L. Temperature program for graphite tube furnace AAS

M. Dissociation continua of sodium salts.
Spectral interference always occurs whenever absorption of radiation from the hollow-cathode lamp is possible due to the overlapping of sharp atomic lines and broad molecular bands of the impurities, scattering of radiation on particles of the impurities which were not vaporized or an indirect influence of the atomization mechanism (see B). This background absorption is caused especially by molecules of gaseous substances and by radiation scattering of particles. Dissociation continua of alkali metal halides with characteristically broad maxima are observed particularly frequently.

N. Background correction. The spectral differences between atomic and molecular absorption provide an opportunity for correcting the non-specific background. In addition to the hollow-cathode lamp, which emits a linear spectrum, a continuum radiator such as the deuterium arc lamp is also used. Using the monochromator, only the resonance line having a bandwidth of 0.002 nm is collimated from the spectrum of the hollow-cathode lamp. On the other hand, a bandwidth of 0.2 to 0.7 nm (see also I) is isolated from the spectrum of the continuum radiator. If they have the same radiation intensity, the radiation from both sources is greatly weakened by broadband background absorption. On the other hand, the very sharp line absorption has hardly any reductive effect on the intensity of the continuum radiator. Hence in this case, a correction takes place directly on the resonance line.

O. Function of a deuterium background corrector. Hollow-cathode lamps and deuterium lamps are arranged so that radiation can be sent through the flame alternately from both radiation sources with a rotating sector mirror. The outlet slit of the monochromator is adjusted such that a maximum of 0.2 to 0.7 nm of the continuum is allowed through.

The intensity of the deuterium lamp is reduced down to the resonance line, but the bandwidth of this line only amounts to approximately 1/100th of the masked continuum, so that even with 100% absorption of the resonance line from the primary radiation source a 1% absorption of the continuous radiation is detected. The intensities of the HCL and the D_2 emitters are equalized to $I_{HCL}/I_{D_2} = 1$. In this way the radiation scattering and molecular absorption, which both reduce radiation to the same degree, are compensated as background. However, if there is an element-specific atomic absorption in addition to the non-specific background absorption, then the intensity I_{HCL} is reduced additionally by the amount of the absorption.

P. The principle behind Zeeman correction. The splitting up of spectral lines (because of the splitting of energy levels in the atoms) in a magnetic field is called the Zeeman effect, after its discoverer. In a strong magnetic field (e.g. 1 T), in addition to the original resonance line, two additional lines appear which are shifted approximately 0.01 nm to shorter or longer wavelengths. There is a polarization of the radiation at the same time. If the magnetic field is at a right-angle to the direction of the radiation, the π-components are polarized in a direction parallel to the magnetic field, whereas the σ-components are polarized perpendicular to the magnetic field (transverse Zeeman effect). To take advantage of this effect, a pulsing magnetic field is applied to the graphite tube furnace; this is called the inverse Zeeman effect, in contrast to the direct effect with a magnetic field on the source of the radiation. The radiation of the primary power source remains unchanged (the resonance line is not split). To measure the background absorption, a polarizer oriented perpendicular to the magnetic field is positioned in the path of the radiation. This blocks the radiation polarized parallel to the magnetic field, and therefore the atomic absorption of the ν-components. Therefore, when the magnetic field is switched on, only the background absorption is measured and not any atomic absorption.

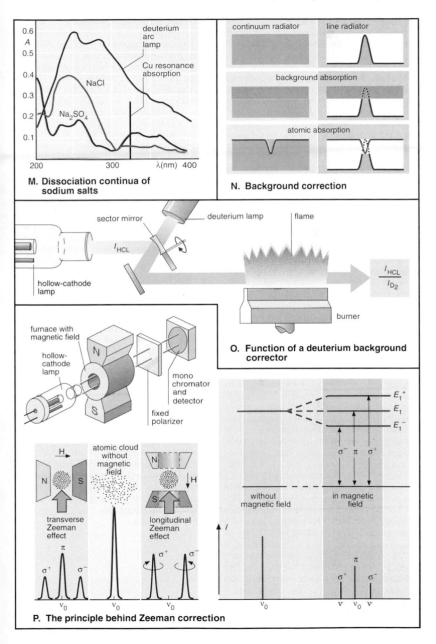

M. Dissociation continua of sodium salts

N. Background correction

O. Function of a deuterium background corrector

P. The principle behind Zeeman correction

Q. Determination of lead in blood. Trace levels of lead in blood samples (or serum samples) are preferentially conducted with flameless AAS because of the low limits of detection. However, by using the boat technique, the flame method with a shorter analytical time can also be applied. According to Delves, a small amount of sample (maximum 1 ml of blood) is placed in a small, round nickel container and the solution is dried (e.g. near the flame of an AAS device or in a muffle furnace). Then the boat is placed in the flame, where it is atomized within a few seconds. One obtains a high, narrow signal, whereby interference due to the matrix is eliminated via the background correction (see the comparison of signals 1 and 2). Temperatures which can be attained with the sample boat method lie around 1200 °C, so only readily atomizable elements such as arsenic, bismuth, lead, cadmium and thallium, all of which have toxic effects, can be determined.

With an atomizer–burner system (see E), only about 10% of the sample is transferred to the flame, and therefore into the absorption space. Except for the determination of lead, generally a tantalum boat is used for this method.

R. Determination of lead in refuse drainage water. Using Zeeman background correction in flameless AAS makes interference-free analysis possible in complex drainage water samples from refuse dumps heavily loaded with organic material. Iso-formation aids prevent the uncontrolled loss of elements via evaporation before atomization. When determining lead, a no-loss pretreatment of the sample in a L'vov platform (see G) is possible only up to 500 °C; however, at this temperature it is difficult to prevent interference, as shown by the signals of the sample and after the addition of a standard solution in (1) in comparison with the standard solution alone with 0.2 ng of Pb. After the addition of 1% phosphoric acid as an isoformation aid, it is possible to pretreat the water sample at 800 °C (2). The addition results in a clear increase in the sample signal; the shift of the sample signal compared with the standard solution signal is still small.

S. Determination of selenium

1. In blood serum with graphite tube furnace AAS. A comparison of signals a and b shows the different effects of background correction using a continuum force (see N) vs the Zeeman effect (see Q): the interference-free analysis of selenium is possible only with the help of the Zeeman effect.

2. In water samples with the hydride method. Using the reducing agent sodium borohydride ($NaBH_4$), only selenium(IV) compounds are reduced to selenium hydride. Selenium is usually present in the $+6$ oxidation state (a) in water samples. Selenium(VI) must first be reduced to selenium(IV) (b) by heating with semi-concentrated HCl. As a rule, with the hydride method a matrix effect which necessitates background correction does not occur.

T. Determination of phosphorus

1. Comparison of graphite tube furnace and L'vov platform. In biological materials, phosphorus determination on the resonance line of 213.6 nm is always subject to interference from large amounts of iron. When a L'vov platform is used (see G), the background absorption is considerably less than in a graphite tube furnace (see H), owing to the very rapid heating rate and atomization.

2. Determination in soyabean oil. Trace elements in oils and fats can often be determined using the graphite tube furnace method with an organic solvent and after the sample has been diluted. However, as a rule, very laborious temperature programs are necessary for interference-free analysis in order to achieve the complete isolation of non-specific background and specific element absorption.

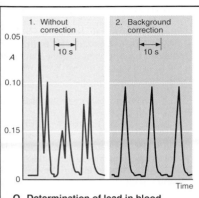

1. Without correction
2. Background correction

Q. Determination of lead in blood

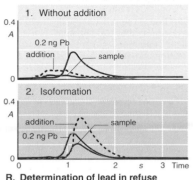

1. Without addition
2. Isoformation

R. Determination of lead in refuse drainage water

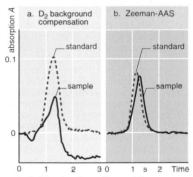

a. D₂ background compensation
b. Zeeman-AAS

1. In blood serium with graphite tube furnace AAS

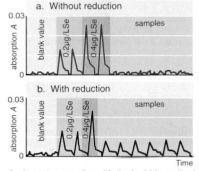

a. Without reduction
b. With reduction

2. In water samples with the hydride method

S. Determination of selenium

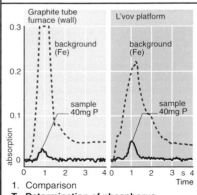

Graphite tube furnace (wall)
L'vov platform

1. Comparison

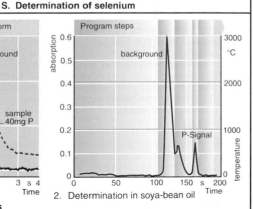

Program steps

2. Determination in soya-bean oil

T. Determination of phosphorus

7.2 Atomic Emission Spectrometry

Definition. The characteristic radiation energy emitted by excited atoms is used in atomic emission spectrometry for the detection and quantitative analysis of elements. A spectral line corresponds to the difference between energy states of an electron. The state of an atom with respect to its electrons can be indicated by four quantum numbers: the primary and secondary quantum numbers n and 1, respectively, the magnetic quantum number m_1 and the spin quantum number m_s. Depending on the wavelength of the emitted radiation, in spectrochemical analysis one differentiates between optical atomic spectra (in the UV/visible range) and X-ray spectra. Further differentiation of atomic emission spectrometry is based on the excitation source.

A. Path of rays in a spectrograph. It has been known for about 100 years that heating (e.g. in a flame or by electric sparks) induces characteristic radiation from metal vapors or noble gases in which individual, free atoms are present, and this radiation can be separated spectrally into individual lines. This spectral separation occurs in a spectrograph by means of a prism or a grid. If one generates a slit along with the light source, one obtains the individual wavelengths in the form of spectral lines—as a line spectrum which can be reproduced on a photographic plate.

B. Energy level diagram and line spectrum of hydrogen. The hydrogen atom produces relatively simple line spectra. They are based on the principle of quantum theory that electrons in an atom can only occupy certain selected energy states. If spectral lines result from transitions with a common atomic level, then they are combined into a series.

1. Energy level diagram. Starting from the lowest energy or ground state, the electrons of the atoms are raised to excited states (i.e. higher energy states) by taking up energy quanta. Lyman series lines are generated when the electrons fall from the various excited states back down to the ground state. Lines which are produced via the transition of electrons to the base state are called resonance lines. Lines of the Balmer or Paschen series are produced in the transition to the second or third energy level, respectively.

2. Line spectrum of hydrogen. According to Planck's equation, $E = E_2 - E_1 = h\nu = h/\lambda$, the lines of the Lyman series exhibit the shortest wavelengths. They lie in the vacuum UV region. The Balmer series produces emission lines in the visible range and the Paschen series in the infrared spectral region.

C. GROTRIAN diagram. Since the number of electron transitions, and therefore the number of lines, are determined by the number and arrangement of the outer electrons of an element, atoms with a small number of outer electrons display spectra with relatively few lines. The GROTRIAN diagram can be used to compile the energy levels which are possible for an atom. The symbols indicate the energy levels: thus, 3 $^2S_{1/2}$ is the ground state of sodium. The resonance line doublet is generated with 589.0 (3 $^2S_{1/2}$–3 $^2P_{3/2}$) and 589.59 nm (3 $^2S_{1/2}$–3 $^2P_{1/2}$). The label $1/2$ indicates a doublet, the letters stand for secondary quantum numbers: S ($l = 0$), P ($l = 1$), D ($l = 2$) and F ($l = 3$). The level of lowest energy is arbitrarily set equal to zero. In addition to the transitions which are permissible according to the selection criteria, the diagram also shows the associated energies in eV and wavelengths of the emitted radiation.

D. Excitation sources. Optical emission spectrometry uses a flame, an electric spark, an arc, plasma and lasers as excitation sources. In X-ray emission spectrometry, X-radiation is added to this list. In a spark both ionization and excitation occur. X-ray beams cause the removal of electrons from an inner electron shell.

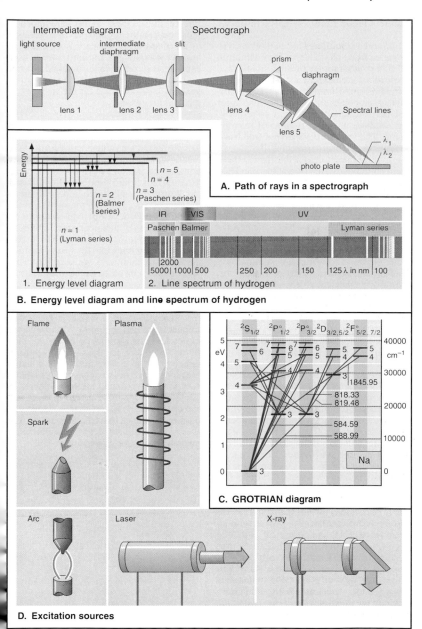

A. Path of rays in a spectrograph

1. Energy level diagram
2. Line spectrum of hydrogen

B. Energy level diagram and line spectrum of hydrogen

C. GROTRIAN diagram

D. Excitation sources

Flame Atomic Emission Spectrometry

A. Flame excitation. This method used to be called flame photometry. The evaluation of flame colorations for the qualitative analysis of elements was first applied by Kirchhoff and Bunsen. In a flame, the following processes occur rapidly one after another: an aqueous solution containing, e.g., sodium chloride, is sprayed into the flame. The solvent is vaporized (a); NaCl remains in the form of particles. Because of the high temperatures in the flame, these solid compounds are also vaporized. They partially dissociate into atoms, according to the equation $NaCl_{(s)} = NaCl_{(g)} = Na_{(g)} + Cl_{(g)}$ (b). Part of the atoms in the gaseous state (g, gaseous; s, solid) are excited by thermal energy: $Na_{(g)} = Na^*_{(g)}$ (c). As a result of the return of the electrons to the ground state, light is emitted—at the main wavelength of 589 nm for sodium (d). Undesirable side reactions are ionizations, which occur particularly readily with alkali metals with their low ionization potentials. The formation of metal oxides is another interferance.

B. Term diagrams and emission lines for lithium. In a Bunsen burner flame, lithium salts give off a crimson red color, which is mostly covered up by the sodium coloration, however. The simplified term diagram (see p. 92, B and C) for lithium produces transitions, the emission radiation of which extends from 274 to 671 nm (1). In a hand spectroscope (see p. 93, A, for the path of the rays), the orange–yellow line at 610.2 nm and the red line at 670.8 nm can be readily distinguished from the sodium line at 589 nm (2).

C. Structure of a flame photometer. A flame photometer (see Section A concerning the term) consists of a burner with an atomizer (see D) and an optical part. The spectral separation of the radiation proceeding from the flame occurs after passing through a slit with the aid of a monochromator (grid or prism). Photoelements are used as detectors. With dual-beam devices, the radiation is split into two parts: in the first part, the intensity of the analytical line of the element to be determined is measured; in the second part, the radiation of an added inner standard is measured at another wavelength. Lithium is often used as an inner standard in this method. Flames with lower temperatures such as the town gas–air flame are used for the easily excitable alkali metals.

D. Different atomizer designs. The atomizer has the task of aspirating the fluid sample and of spraying it in the form of the smallest possible droplets into the flame at a constant rate. Using a gas flow under high pressure, the liquid flow is (a) divided into small droplets and (b) fed into the flame as an aerosol. As a result of the suction of the gas flow, a mist develops in the atomizer jet with droplets of different sizes. Droplets larger than 20 μm (c) do not reach the flame, but rather flow back into the spray chamber. These direct atomizers are used for turbulent flames (with gases having a high rate of combustion, such as H_2–O_2).

For laminar flames (without many vortices), the larger droplets are captured after the aerosolization chamber (d) on the impact plate and are removed through the outlet. Only the small droplets arrive at the burner with the air flow. Since approximately 95% of an aspirated solution leaves the atomizer chamber again in this manner, the efficiency of this atomizer device is considerably less than that of a direct atomizer (a) for turbulent flames. However, atomizers with an impact plate have proven themselves with respect to reproducibility and uniform mode of operation.

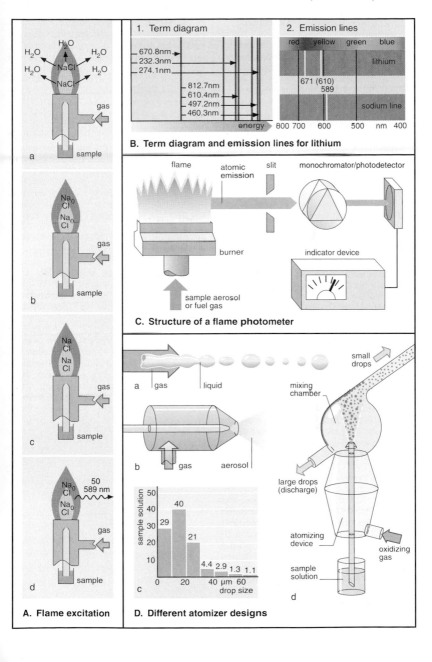

A. Flame excitation

B. Term diagram and emission lines for lithium

1. Term diagram
   - 670.8nm
   - 232.3nm
   - 274.1nm
   - 812.7nm
   - 610.4nm
   - 497.2nm
   - 460.3nm

   energy

2. Emission lines
   red | yellow | green | blue
   lithium
   671 (610)
   589
   sodium line
   800 700   600   500   nm 400

C. Structure of a flame photometer

flame — atomic emission — slit — monochromator/photodetector — burner — indicator device — sample aerosol or fuel gas

D. Different atomizer designs

a gas liquid — small drops
b gas aerosol — mixing chamber — large drops (discharge) — atomizing device — oxidizing gas — sample solution
c sample solution / drop size: 29, 40, 21, 4.4, 2.9, 1.3, 1.1 — 0 20 40 µm 60

Emission Spectrometry with Plasma Excitation

ICP emission spectrometry is a multielement analytical method which has considerably fewer matrix effects than atomic absorption spectrometry (AAS) but with comparable detection limits.

A. Different plasma burners. An atomizer (see also p. 94) is used to produce an aerosol from the sample solution, and argon is added to the aerosol. In this form of very fine droplets, the mixture reaches the actual plasma—generally a gas, the atoms or molecules of which are partially dissociated into ions and electrons.

1. Inductively coupled high-frequency plasma (ICP). In ICP, the charged particles are generated by ionization in the induction coil of a high-frequency generator (HF coil) which is wrapped around the quartz tube through which the material flows. Argon is the actual plasma gas, which can easily be ionized. The plasma burner consists of three concentric quartz tubes. In the innermost tube, the sample aerosol is transported in the argon flow into the plasma; argon (as an auxiliary gas with slight flux) is introduced in the middle tube; and the plasma–argon is introduced in the outer tube. The middle, tulip-shaped tube causes congestion and subsequently rapid acceleration of the plasma–argon along the inside of the outer tube. A ring-shaped plasma is formed, and carrier gas and sample aerosol penetrate axially into the plasma. In the center of the plasma, a tunnel develops having temperatures of 6000 to 8000 K.

2. Direct current plasma (DCP). This is produced in an electric direct current arc which is formed between the carbon and metal electrons (a tungsten electrode in the figure). The three-electrode arrangement enables a suitable form of the plasma to develop which makes possible the optimal supply of the sample aerosol. Temperatures between 5000 and 7000 K are reached.

ICP and DCP equipment has seen the most widespread analytical applications; microwave plasmas have not caught on to the same extent.

B. Simultaneous spectrometer. In contrast to sequential multielement determination, which might use a monochromator driven by a stepper motor, a simultaneous spectrometer uses a Rowland circle in a Paschen–Runge arrangement. Spectral dispersion of polychromatic radiation and focusing can be achieved simultaneously through a grating scratched in a concave surface. The focusing takes place on the Rowland circle (with a radius which is half the curvature of the grating). A focusing lens is not necessary here. Up to 48 elements in a sample can be determined within a few seconds in a simultaneous spectrometer. On a Rowland circle, there is room for a maximum of 48 slits and for photomultipliers suitable for the corresponding spectral regions. Background correction can also be effected by moving the slit and thereby slightly deflecting the light.

C. Resolution of the iron triplet. The example with the resolution of iron lines with spacings of fractions of nanometers shows how powerful an ICP spectrometer is.

D. Cu and P determination in waste water. An example of an application in environmental analysis is the simultaneous determination of traces of copper and phosphorus (phosphate) in waste water, where here too the spectral lines have a spacing of only 0.02 nm.

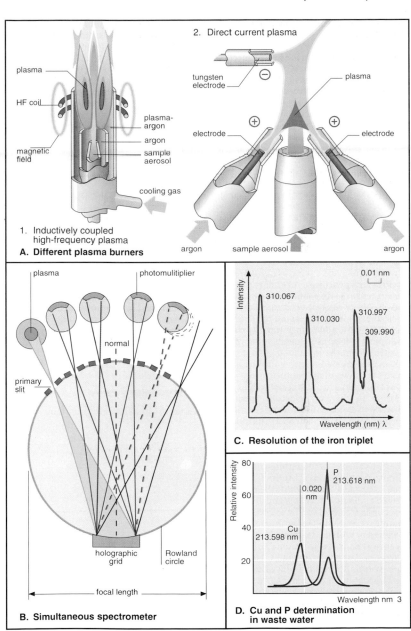

2. Direct current plasma

plasma

tungsten electrode

plasma

electrode

electrode

argon   sample aerosol   argon

plasma

HF coil

magnetic field

plasma-argon

argon

sample aerosol

cooling gas

1. Inductively coupled high-frequency plasma

A. Different plasma burners

plasma   photomultiplier

normal

primary slit

holographic grid   Rowland circle

focal length

B. Simultaneous spectrometer

0.01 nm

Intensity

310.067

310.030

310.997

309.990

Wavelength (nm) λ

C. Resolution of the iron triplet

Relative intensity

80

60

40

20

0.020 nm

P 213.618 nm

Cu 213.598 nm

Wavelength nm 3

D. Cu and P determination in waste water

7.3 X-ray Fluorescence Analysis (XRFA)

A. Term transitions in the inner electron shells. If very high energy radiation (0.1 to 100 keV) hits atoms having higher atomic numbers, electrons can be expelled from the inner shells. The vacancies that are created are filled by electrons from shells lying further out. This process generates spectral lines called X-rays. If an electron is knocked out of the innermost, or K shell, then one observes the K spectrum. The corresponding emitted X-ray lines are combined into series according to the shells or the energy levels, as well as the possible term transitions between the K, L and M shells. The transitions shown with lines are not allowed based on the selection rules, according to quantum theory (n, primary quantum number; l, secondary quantum number; m, magnetic quantum number).

B. Primary processes. A primary process that occurs in the interaction between atoms and X-rays or electrons is absorption with photoionization of the atom (1). The corresponding absorption spectrum (2) shows absorption edges as a specific absorption resulting from the ionization. The dependence of the emitted electrons on the kinetic energy is shown in the photoelectron spectrum (3), where the absorption edges and maxima in the photoelectron spectrum correspond to one another. Both characterize particular energy levels of individual atoms (see A).

C. Secondary processes. If one assumes the equivalence of electromagnetic and corpuscular radiation (i.e. from X-rays and from electrons), then both of them induce the same primary processes and also occur as secondary radiation. The excited state of an atom is canceled either by the emission of X-radiation with the resultant X-ray fluorescence or emission spectrum (2), or by the release of Auger electrons (1). The X-ray fluorescence spectrum is invoked by the appearance of X-ray quanta, while the Auger electron spectrum comes as a result of the Auger or inner photoelectric effect.

D. The principle behind the excitation of X-ray fluorescence radiation. Electron, ion, X- or gamma radiation can be used as high-energy radiation. In every case a characteristic X-radiation occurs as a result of the removal ('knocking out') of electrons from inner electron shells and the refilling of the vacancies by electrons from further outlying shells. To demonstrate the process, an atomic envelope with electron shells K, L, M and N was used as a model. If the energy hv of the excitation radiation is greater than the binding energy hv_K of an electron e_K in the K shell, then this electron will leave the electron shell with a transfer of energy. Fluorescence radiation which is characteristic for an atom appears. If, for example, the K_α fluorescence radiation causes the removal of an electron from the M shell (e_M), then the Auger effect and therefore a double ionization have been introduced. The Auger effect represents a competing reaction for the fluorescence emission. The fluorescence yield, as the ratio of the number of emitted fluorescence radiation quanta to the number of vacancies generated in the same amount of time, decreases. For heavy elements, the fluorescence yield amounts to approximately 70–90%. According to Moseley's law, the wavelength of a K_α radiation depends on the atomic number Z of the relevant element:

$$1/\lambda = \text{constant}(Z - s^*)^2$$

where s^* is the screening constant.

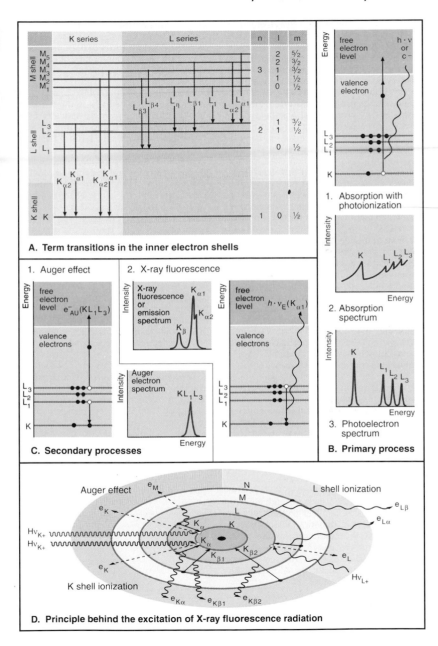

A. Term transitions in the inner electron shells

B. Primary process

1. Absorption with photoionization
2. Absorption spectrum
3. Photoelectron spectrum

C. Secondary processes

1. Auger effect
2. X-ray fluorescence

D. Principle behind the excitation of X-ray fluorescence radiation

E. Energy- and wavelength-dispersive XRFA. In energy-dispersive XRFA (EDXRFA), the emitted fluorescence radiation is dispersed according to its energy using a semiconductor detector (the analyzer crystal); an example is shown for three different wavelengths 1, 2 and 3. Radiation quanta are converted directly into energy-proportional voltage impulses. In wavelength-dispersive XRFA (WDXRFA), the fluorescence radiation is separated according to wavelength by diffraction on an analyzer crystal (with known interplanar spacing given different goniometer angles according to Bragg's law). Wavelength-dispersive spectrometers have higher resolution, but energy-dispersive spectrometers have higher sensitivity.

F. The principle behind an X-ray tube. When a high voltage U_a is applied, the electrons that are released due to the thermionic emission from a metal cathode in a vacuum tube are accelerated. They fall on the anode or anti-cathode, where they are slowed down. In addition to heat, X-radiation is generated, with wavelengths between 10^{-8} and 10^{-11} m (10 and 0.01 nm) (1). The characteristic emission lines (shown here for rhodium) superimpose the continuous spectrum (2).

G. Construction of X-ray spectrometers. In EDXRFA, the detector system consists merely of a solidly mounted semiconductor detector, in which the X-ray fluorescence radiation is demonstrated by an energy loss corresponding to its energy and, following amplification of the signals in a multichannel analyzer, is shown as a spectrum arranged according to its energy.

WDXRFA requires more apparatus. The emitted X-ray beam goes through a collimator and hits an analyzer crystal, through which only that radiation which has a wavelength corresponding to the adjustable goniometer angle according to the Bragg equation is detected. The different wavelengths are determined by adjusting the angle of the goniometer.

H. Different energy resolution. WDXRFA has a greater resolution (see also

E), as shown by the example of the manganese K_α and K_β lines. It is also obvious that a considerably lower radiation intensity is detected with a counter tube or scintillation detector (see I).

I. Detectors. *1. Scintillation counter.* An NaI single crystal (thallium-doped) in a light-proof arrangement is optically coupled to a photoelectronic multiplier. The photoelectric effect is the basis of the measurement: as a result of the absorbed X-ray quanta, photons are produced which reach the photocathode and generate secondary electrons there. Between the photocathode and the multiplier's anode, there is a high voltage (up to 1000 V), divided m-fold by a chain of resistors. Owing to the moving secondary electrons, two to four electrons each will be expelled from m dynodes (multiplication).

2. Counter tube. Electrons generated by the photoelectric effect produce secondary ions and electrons in a gas (e.g. argon). Because of the strong electric field near a thin wire, the electrons reach such a high energy that impact ionization causes a large number of ions and electrons to be formed, which can be counted.

3. Semiconductor detector. The working principle of an Si(Li) detector lies in the p–n transition (semiconductor principle). At a voltage of 1000 V, no free charge carriers are present in the detector volume. The incident radiation ionizes the detector material (photoelectric effect). The resulting charge carriers are drawn away in the electric field, and the current impulse is proportional to the energy converted in the semiconductor.

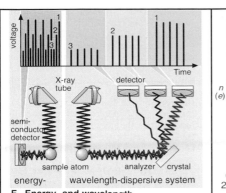

E. Energy- and wavelength-dispersive XRFA

energy-   wavelength-dispersive system

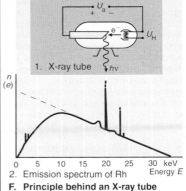

1. X-ray tube

2. Emission spectrum of Rh

F. Principle behind an X-ray tube

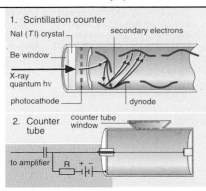

G. Construction of X-ray spectrometers

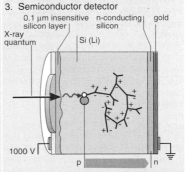

H. Different energy resolution

1. Scintillation counter

NaI (Tl) crystal
Be window
X-ray quantum hν
photocathode
secondary electrons
dynode

2. Counter tube

counter tube window
to amplifier
R

3. Semiconductor detector

0.1 μm insensitive silicon layer
n-conducting silicon
gold
X-ray quantum
Si (Li)
1000 V
p   n

I. Detectors

J. Sample preparation. The most important industrial applications of XRFA are in the iron and steel industry, in cement manufacture, in the feed and fertilizer industry and generally in the metalworking industry. To perform the analyses usually one uses solids in which multielement analyses can be performed non-destructively. The sample preparation is very important here also.

With metals, non-iron alloys and steels, the surface of the sample must be mechanically ground and polished. With powder-like materials that are pressed into tablet form, such as natural stones, ores, minerals, glass types, cement and plastics, the particle size determines the homogeneity of the sample surface. With respect to the accuracy and precision of the analyses, the surface of the pressed tablets must be extremely smooth and homogeneous. With rock samples, a grinding stock with particle diameters of less than $50 \mu m$ is produced. To produce powder tablets, the finely ground powder is directly pressed with binding agents such as methylcellulose, wax or an organic polymerization agent. Dissolving tablets are produced using lithium borate as a fluxing agent. In environmental analysis, films impregnated with paraffin, silicone oil, etc., are used as air filters for the analysis of air particles or suspended material in gases (aerosols). Oils can be analyzed directly. Aqueous solutions can be concentrated by evaporation on quartz carriers.

K. EDXRFA analysis of a steel alloy. With EDXRFA, optimal excitation conditions can be achieved for elements from phosphorus ($Z = 15$) to neodymium ($Z = 60$) using modern heavy-duty tubes. The evaluation is done using standard or reference samples. The elements Mg, Al, Si, S, Sn, Ti, V, Cr, Mn, Ni and Cu are in the foreground of the analysis. Rhodium was used as the anode here; its signal provides no noise. The measuring time is $200 s$ at an energy level of $0.2 keV$, an anode voltage of $10 kV$ and an anode current of $50 \mu A$. Portable EDXRFA systems with a radioactive excitation source are available for this non-destructive analysis of metallic workpieces.

L. WDXRFA analysis of a granite sample. From the recorded diagram of a WDXRFA analysis, in addition to qualitative statements about the presence of particular elements, one can also obtain semiquantitative data (with relative standard deviations around 30%). The spectrum of a standard granite sample also shows the superposition of lines (Zr/Sr and Rb/Y). Because of the spectrum which otherwise has few lines, the amounts of Nb, Th, Zr, Rb, Ba and Sr can be determined in the range between 20 and 250 ppm with this method.

M. EDXRFA analysis of a dust filter. Suspended dust analyses are conducted in environmental research with the goal of determining particular emitters. Prerequisites for XRFA analysis are small and constant blank values of the empty filter, a reproducible geometric distribution of the collected dust particles on or in the filter and a high ratio of mass accumulation to dead weight. After appropriate calibration, good resolution makes possible the quantitative analysis of Ti, V, Cr, Ba, Mn, Ni, Cu, As and Se with the area covered being between 43 (Cr) and 1460 (Cu) ng/cm^2. Because of the interference with Ti–Ba and As–Pb, with these elements there is a relatively high systematic error rate of more than 20%.

N. EDXRFA analysis of an oil sample. To determine sulfur rapidly using WDXRFA, pressed tablets are formed, as with powders. With EDXFRA, even trace amounts of metals can be determined without laborious sample preparation in order to characterize the origin of the heavy metal grindings (with machine oils). With an anode voltage of $30 kV$, Ti, V, Cr, Mn, Fe, Ni, Cu, Zn, Mo and Pb can be measured within $400 s$.

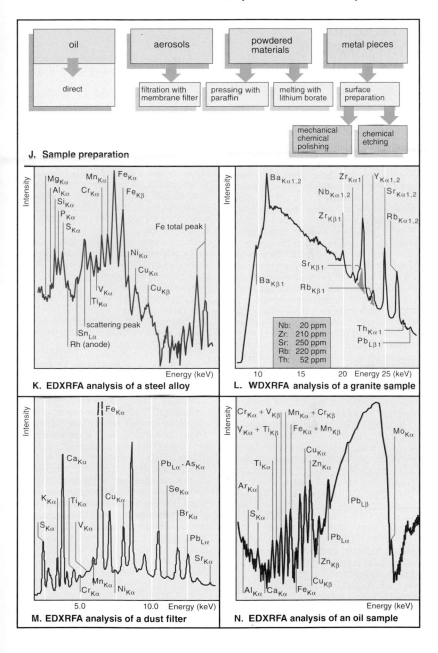

J. Sample preparation

K. EDXRFA analysis of a steel alloy

L. WDXRFA analysis of a granite sample

Nb: 20 ppm
Zr: 210 ppm
Sr: 250 ppm
Rb: 220 ppm
Th: 52 ppm

M. EDXRFA analysis of a dust filter

N. EDXRFA analysis of an oil sample

8 Molecular Spectroscopic Analytical Methods

8.1 Introduction

A. Electromagnetic spectrum. Light is defined physically as a transverse electromagnetic wave motion with periodic changes of electric and magnetic fields. According to the relationship which Planck described between the energy of a photon and the frequency ($E = hv$, where h = Planck's constant), electromagnetic waves differ in their frequency v or wavelength λ (with $v = c/\lambda$ where c is the speed of light) and therefore also in the energy of their photons (light quanta). Short wavelengths correspond to high-energy radiation and long wavelengths to low-energy radiation. The following values are used to characterize an electromagnetic wave: the wavenumber is the reciprocal wavelength: $= 1/\lambda$ in cm^{-1}. The frequency v indicates the number of vibrations of an electric or magnetic field per unit time [vibrations per second in s^{-1} or Hz (hertz); $1 \text{ MHz} = 10^6 \text{ Hz}$]. The velocity of propagation in a vacuum is the same for all electromagnetic waves with $c = 2.997925 \times 10^{10} \text{ cm/s}$. The wavelength λ can be derived from the equation $c = \lambda v$ as c/v.

In actual practice, the electromagnetic spectrum is divided into different spectral regions: radio waves have wavelengths greater than 0.1 m and microwaves between 500 μm and 30 cm; the infrared region is divided into far-, mid- and near-infrared. The visible (VIS) region includes wavelengths from 400 to 800 nm and the near-ultraviolet (UV) and vacuum UV regions extend from 400 to 200 or 100 nm ($1 \text{ nm} = 10^{-9} \text{ cm}$). Molecular spectroscopic analysis methods are based on interactions between molecules and light quanta with different energy levels. They are named after the spectral region and imply different effects on the molecules, depending on the energy, effects such as electron excitation, molecular vibration and molecular rotation.

B. Electron excitation and molecular vibration. The absorption of UV or visible light by organic molecules leads to the excitation of π-electrons (of double and triple bonds) or n-electrons (non-binding electron pairs of oxygen, sulfur and nitrogen atoms). On the other hand, the energy of this spectral region is not sufficient to excite σ-electrons (of the molecular backbone). The energy uptake from the absorbed light quanta between 160 and 8000 kJ/mol leads to excitation of the electrons from the ground state (with the energy level E_1) to higher energy levels (E_2). Electron transitions are generally shown graphically in a term diagram.

IR radiation (low energy level) causes molecular vibration, i.e. mechanical vibration between the atoms of a molecule. As the example with the carbonyl group shows, the vibration can occur in the direction of the binding-valence axis as valence vibrations, or with the deformation of the binding angle they appear as deformation vibrations. The energy levels of IR radiation lie between 160 and 2400 kJ/mol.

C. Absorption and emission of light by molecules. If irradiated light having an intensity I_0 is partially absorbed by molecules in a solvent (in a cuvette having a layer thickness d), a photometer can be used to measure the residual (transmitted) intensity I as absorption or transmission T ($A = \log 1/T$). If the portion of irradiated light intensity absorbed by the molecules is recorded as a function of the wavelength, an absorption spectrum is produced (in the UV–VIS region: electron-excitation spectrum). The energy taken up is released again as heat when the electrons undergo transition back to the ground state. On the other hand, if light (radiation energy) is emitted again as a result of the irradiated light, one obtains an emission (fluorescence) spectrum using a fluorimeter (see p. 112).

Type of transfer			molecular rotation	molecular oscillation	electron excitation
Spectroscopic methods		microwave absorption	infrared spectroscopy		UV–VIS spectroscopy
Spectral region	radio waves	microwaves	infrared (IR)		Vis UV	X-rays
			F	M	N	

wavelength [m] $\lambda = \dfrac{c}{\nu}$ λ	10	1	10^{-1}	10^{-2} 1cm	10^{-3}	10^{-4}	10^{-5}	10^{-6} 1µm	10^{-7}	10^{-8}	10^{-9} 1nm	10^{-10} 1Å
Frequency [Hz] $\nu = \dfrac{c}{\lambda} = c \cdot \tilde{\nu}$ ν	10^{-7}		10^{-9}		10^{-11}		10^{-13}		10^{-15}		10^{-17}	
Wavenumber $\tilde{\nu}$ $\tilde{\nu} = \dfrac{1}{\lambda} = \dfrac{\nu}{c}$ [cm^{-1}]	10^{-3}		10^{-1}		10		10^{3}		10^{5}		10^{7}	

A. Electromagnetic spectrum

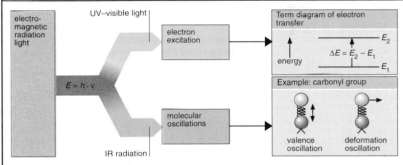

B. Electron excitation and molecular vibration

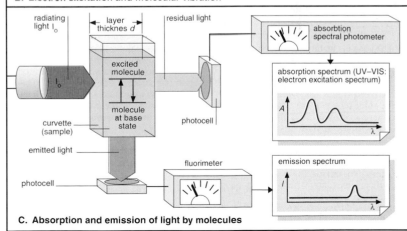

C. Absorption and emission of light by molecules

8.2 Colorimetry

Basic principles. In general, colorimetry today includes all those physical-chemical methods which make use of the light absorption of colored solutions for the quantitative analysis of dissolved substances. In a more restricted sense, colorimetry is based on a direct color comparison without any measuring device. In contrast to spectrophotometry, colorimetric measurements use simple equipment and undispersed light. When electromagnetic radiation (of a light beam) passes through a solution, its intensity is reduced due to absorption. Bouguer (1698–1758) and Lambert (1728–1777) found that the absorption A is dependent on the layer thickness d of the measuring cuvette: $A = k_1 d$. Beer (1825–1863) determined the proportionality of the absorption with a constant layer thickness solely from the molar concentration c of a light-absorbing material: $A = k_2 c$ (k = coefficient). The combination of both equalities yields the Bouger–Lambert–Beer law, usually called the Lambert–Beer law: $A = \varepsilon(\lambda)cd$ (ε is the molar or spectral absorption coefficient in $L\,mol^{-1}\,cm^{-1}$, d is the layer thickness in cm and c is the concentration in mol/L). As per DIN 1349, the formerly common term extinction E ($= \log I_0/I$; I_0 is the initial intensity and I is the transmitted intensity) was replaced by absorption A and the molar extinction coefficient was replaced by the molar absorption coefficient.

A. Dubosq's immersion colorimeter. Colorimeters are devices with which the measurement of the absorption occurs based on the Lambert–Beer law using undispersed light. In Dubosq's (1817–1886) colorimeter, there is a cuvette with the reference solution on the right and a cuvette with the analytical solution on the left. Massive glass cylinders are immersed in both fluids, the immersion depths of which are compared thus: after properly adjusting the mirror (below), one looks through a lens (above) and sees a circle bisected by a narrow line; the two semi-circles should have exactly the same shade of color or the same degree of brightness.

The layer thicknesses d_1 and d_2 are read from two scales on the rear side of the instrument. For the unknown concentration c_2, we can derive $c_2 = c_1 d_1/d_2$.

B. Color comparison in the cuvette test. Colorimetry is being used once again as a direct color comparison method in an easy-to-use form, especially in water analysis for quick test procedures. In the cuvette test, a color scale is placed next to the Plexiglas cuvette, in which the sample and necessary reagents are located. This system, called a comparator, is held against the light (artificial light) at eye level for evaluation, and the color of the sample solution is assigned to a reference color.

C. Turntable comparator. Even a water sample's natural color can be compensated for by using two cuvettes, one cuvette for the untreated analytical sample. By turning the color wheel in front of the water sample lacking any reagents, one looks for the setting at which a comparable color effect is seen through both windows.

D. Color scale sliding comparator. Visual colorimetry using the principle of remission measurement (direct illumination) is usually applied in color card tests when the layer thickness is increased and therefore the sensitivity is greater. Colors which are valid for a given concentration range and which are located on a plastic color card are compared with the color of a water sample after reagents have been added. In the color scale sliding comparator, the glass with the sample and reagents is situated above a white surface, and the color scale consisting of separate circles is beneath the untreated sample. The scale is slid through until the color effects for both solutions are comparable when viewed from above.

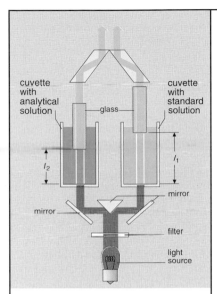

A. Dubsoq's immersion colorimeter

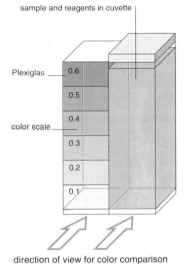

B. Color comparison in the cuvette test

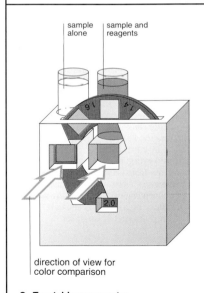

C. Turntable comparator

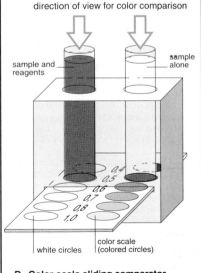

D. Color scale sliding comparator

8.3 Spectrophotometry

A. Visible range spectrometry. The visible light spectrum extends from approximately 380 to 750 nm. A mixture of all wavelengths produces the color white. Prisms or grids can be used to generate radiation of the listed colors with particular wavelength: red 650–750, orange 595–650, yellow 560–595, green 490–560, blue 435–490 and violet 380–435 nm.

B. Wavelength-dependent absorption. A solution always shows the color of the predominant residual light, or complementary color. Red solutions absorb light in the region around 500 nm (green), blue solutions above 600 nm (orange to red) and yellow solutions around 450 nm (blue). Light which displays only a narrow wavelength region is called monochromatic, whereas white light or light from several wavelength ranges (colors) is called polychromatic (see Colorimetry, p. 106).

C. Structure of photometers. A photometer to measure light absorption or transmission basically consists of a light source, lenses and slits for directing the light and a part for generating particular wavelengths.

1. Filter photometer. By using a continuous light source (incandescent lamp as a continuous emitter), these devices are used to select particular wavelengths with the aid of filters. For example, for an orange-colored solution, a filter with the complementary color blue is used, which allows the main wavelength (about 450 nm here) as well as neighboring wavelengths through, although with lower intensity (see also B).

2. Spectrophotometer. A spectrophotometer is used to generate practically monochromatic light in both the visible and UV ranges using monochromators (grids or prisms). The automatic recording of spectra is also often possible with these devices. To compensate for the individual absorption of the solvent, in a single-beam photometer the solvent's absorption is first compensated to zero and then the sample solution is measured. In dual-beam photometers, the light beam can be directed alternately through the reference cuvette (solvent) and measuring cuvette (sample) or separated into reference and measuring beams using a rotating disc.

D. Sources of error when beams pass through a measuring cuvette. Photometric cuvettes must demonstrate high optical quality (quartz in the UV region) and exact dimensions, especially of d. Erroneous absorption measurements can result from light scattering on dirty cuvette windows, from dust particles or clouding in the solution, by the reflection of light on the wall of the cuvette and by solvent absorption. Additional sources of error are temperature effects, incorrect selection of the wavelength setting and insufficient resolving power of the equipment.

E. Transmission and absorption. In a transmission spectrum, one determines the transmission T of a solution as a percentage of the incident light as a function of the wavelength. In an absorption spectrum, the negative logarithm of the transmission is drawn as the absorption A ($A = -\log T$, where $T = I/I_0$) according to the Lambert–Beer law (see p. 106).

F. UV spectrum of a disubstituted benzene derivative. Absorption maxima and the molar (spectral) absorption coefficients (with a known molar concentration) are characteristics of an absorption spectrum. Parts of a molecule which absorb light in the UV–visible range are called chromophores or chromophoric systems, shown here for an aromatic system (with π-electrons) bearing a carbonyl group and an amino group (with n-electrons from oxygen or nitrogen). The substituents of benzene affect the position of the absorption maxima and the size of ε. The highest maximum is especially well suited for photometric analysis.

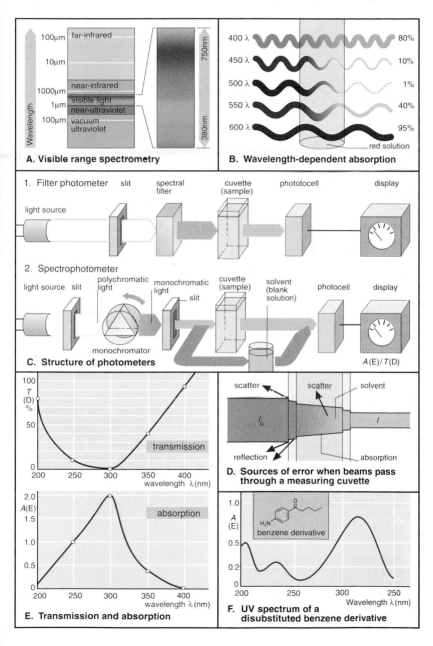

A. Visible range spectrometry

B. Wavelength-dependent absorption

1. Filter photometer

light source — slit — spectral filter — cuvette (sample) — phototocell — display

2. Spectrophotometer

light source — slit — polychromatic light — monochromator — monochromatic light — slit — cuvette (sample) — solvent (blank solution) — photocell — display

A(E) / T(D)

C. Structure of photometers

D. Sources of error when beams pass through a measuring cuvette

scatter — scatter — solvent — reflection — absorption — I_0 — I

E. Transmission and absorption

transmission

absorption

F. UV spectrum of a disubstituted benzene derivative

benzene derivative

G. Effect of the spectral bandwidth.
Radiation of a particular wavelength region λ passes through the slit in a monochromator (see C), which can be a filter, a prism or a grid. The intensity distribution can be portrayed in the form of an isosceles triangle. According to the equation $\lambda = s \, d\lambda/ds$, the spectral bandwidth is specified by the width of the monochromator slit s (approximately 0.1 mm) and the dispersion of the monochromator $d\lambda/ds$ with a specific change in the width ds of the slit. Thus, with a prescribed, device-dependent dispersion, a reduction of the spectral bandwidth can be achieved by narrowing the slit. A narrow slit width is necessary in order to obtain light which is as monochromatic as possible and therefore for the Lambert–Beer law to be valid. However, if the slit is made too narrow, the intensity of the light will be too weak (see triangle distribution). The spectral bandwidth affects the dispersion of an absorption spectrum, as shown by the example for benzene.

H. UV spectra of anthracene and benzene. Alkenes, alkynes and aromatic compounds represent chromophores made of π-electrons. Of benzene's three absorption bands (184, 198 and 255 nm), for measurements one always uses the finely structured band with the longest wavelength which dissolves best in the vapor state. With benzene derivatives which have substituents with n or π-electrons in conjugation with an aromatic system, there is a shift of the absorption maxima to longer wavelengths—a bathochromic effect. This is associated with an increase in the intensity, i.e. of the absorption coefficient, called a hyperchromic effect. Thus, the transition of benzene to anthracene, a polycyclic aromatic, leads to a bathochromic shift and to an increase in the absorption at the maxima (shown here with different absorption scales).

I. Photometry of lanthanum with arsenazo. Non-absorbing materials such as metal ions can be determined photometrically after reaction with a reagent to give a colored metal complex. A prerequisite for this is a sufficient difference in

the absorption spectra between the reagent and the metal–reagent complex. Since the absorptions at the maximum of the lanthanum–arsenazo complexes of reagent (a) and complex (b) overlap, in this photometric procedure it is necessary to measure a reagent blank solution (c).

J. Photometric two-component analysis.
Because of the additivity of absorptions, if there are sufficient differences in the absorption spectra of individual substances, a photometric dual (multi-)component analysis is also possible in a solution of two (or more) substances. The measurements are made at the location of the two maxima. If the absorption coefficients are known for both substances at λ_1 and λ_2, the two equations can be used to determine the concentrations c_I and c_{II} with only two measurements.

K. Photometric calibration function.
Deviations from linearity, i.e. from the Lambert–Beer law, occur as a result of chemical changes (e.g. dissociation, association and interactions of the light-absorbing particles with the solvent—'true' deviations) or physical changes (e.g. scattered light—'apparent' deviations). By setting up a calibration function, the linear measuring range is determined at the same time.

L. Distribution of relative measuring errors. The distribution function for relative errors σ_A/A as a function of the measured absorption A (and thereby of the concentration to be determined) shows a minimum for absorption between 0.2 and about 1.0—the optimum measuring range in which reliable values can be determined in practice (σ_T = absolute error of transmission).

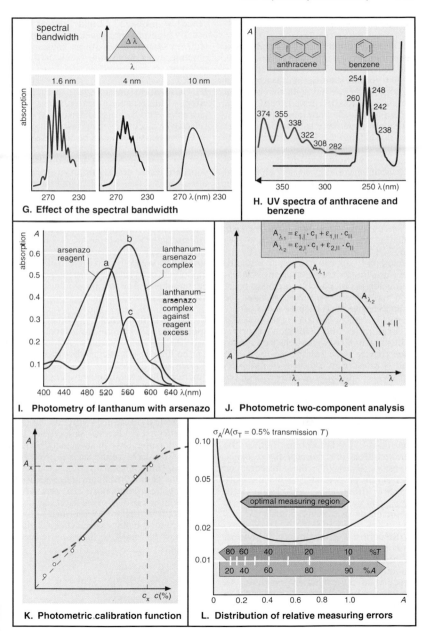

spectral bandwidth

$\Delta\lambda$

1.6 nm   4 nm   10 nm

absorption

270   230   270   230   270 λ(nm) 230

G. Effect of the spectral bandwidth

A

anthracene   benzene

374 355 338 322 308 282   254 260 248 242 238

350   300   250 λ(nm)

H. UV spectra of anthracene and benzene

A

absorption

arsenazo reagent   lanthanum–arsenazo complex

a   b   c

lanthanum–arsenazo complex against reagent excess

0.6
0.5
0.4
0.3
0.2
0.1

400 440 480 520 560 600 640 λ(nm)

I. Photometry of lanthanum with arsenazo

$$A_{\lambda_1} = \varepsilon_{1,I} \cdot c_I + \varepsilon_{1,II} \cdot c_{II}$$
$$A_{\lambda_2} = \varepsilon_{2,I} \cdot c_I + \varepsilon_{2,II} \cdot c_{II}$$

A_{λ_1} A_{λ_2} I + II   II   I

A

λ_1 λ_2 λ

J. Photometric two-component analysis

A

A_x

c_x $c(\%)$

K. Photometric calibration function

$\sigma_A/A (\sigma_T = 0.5\%$ transmission $T)$

0.10

0.05

optimal measuring region

0.02

80 60 40   20   10 %T

0.01

20 40 60   80   90 %A

0   0.2   0.4   0.6   0.8   1.0 A

L. Distribution of relative measuring errors

8.4 Fluorimetry

A. Term diagram (simplified Jablonski diagram). Fluorescence (emission) radiation occurs when electrons pass from the singlet ground state S_0 by absorption of photons to an excited state S_1 (with the vibrational levels 0, 1, 2, etc.), then fall back radiationless to the vibration state 0 from S_1, and return from there back to S_0 with the emission of fluorescent light. There is a nonradiative deactivation in absorption spectroscopy: electronic energy is converted into vibrational energy (A) via an internal conversion (IC). Furthermore, a transition from S_1 to the triplet state T_1 via interspin crossing (now parallel spins of both electrons) is possible, with the subsequent loss of energy in the form of phosphorescence radiation (P). The transition from S_1 to T_1 is called 'intersystem crossing' (ISC). Fluorescence only appears during or immediately after the excitation; because of the S_1–T_1 transition, phosphorescence takes more time and can still be measured after the excitation source has been switched off.

B. Excitation and emission spectra. According to Stokes' law, electron transitions require more energy for excitation than is released in the form of radiation energy. Therefore, the fluorescence spectrum is shifted to longer wavelengths than the absorption spectrum. If the electrons from S_1 return from vibration level 0 to S_0 (0), the level of symmetry is between the excitation and fluorescence spectrum (0–0 transition). The absorption follows the Lambert–Beer law. For the fluorescence intensity $I = \varepsilon I_0 Q K$; the fluorescence intensity is proportional to the intensity of the irradiated (absorbed) light (Q = quantum efficiency, ε = absorption coefficient and K = device constant).

C. Structure of a fluorimeter. In contrast to photometry, in a fluorimeter the radiation intensity of the fluorescence (emission) radiation is measured perpendicular to the direction of the excitation radiation. An emission monochromator is connected in series to remove remnants of the excitation light (scattered light) before the fluorescence is measured. Light sources are usually high-pressure gas discharge lamps with a radiation intensity as constant as possible, and a photomultiplier is used as a receiver.

D. Calibration functions with quenching effect. Fluorescence measurements are characterized by high sensitivity and selectivity. According to the law mentioned in C, the sensitivity can be increased by increasing the excitation intensity (in contrast to photometry). At higher material concentrations, deviations from linearity often appear, since the quantum efficiency Q is reduced here due to self-absorption (transfer of electronic energy from one molecule to another instead of the release of fluorescence radiation) and quenching effects (general lowering of Q). In both cases the electron transitions are obstructed, and radiation energy is lost due to effects of the solvent or by release to the molecules to be measured. The region of linearity is narrower than in photometry.

E. Three-dimensional fluorescence spectrum. Excitation and fluorescence wavelengths and the fluorescence intensity can be linked with one another in a three-dimensional fluorescence spectrum. In practice, the excitation wavelengths are kept constant and the fluorescence (emission) spectrum is recorded.

F. Chemical substances with fluorescent properties. Fluorescent properties are observed particularly in substances with a rigid molecular structure such as aromatic systems, compounds with conjugated double bonds, carbonyl linkages and condensed heterocyclic compounds. The part of a molecule responsible for the fluorescence is called a fluorophore. With metal ions, fluorescence labeling is used with a chelating agent for sensitive fluorimetric measurements.

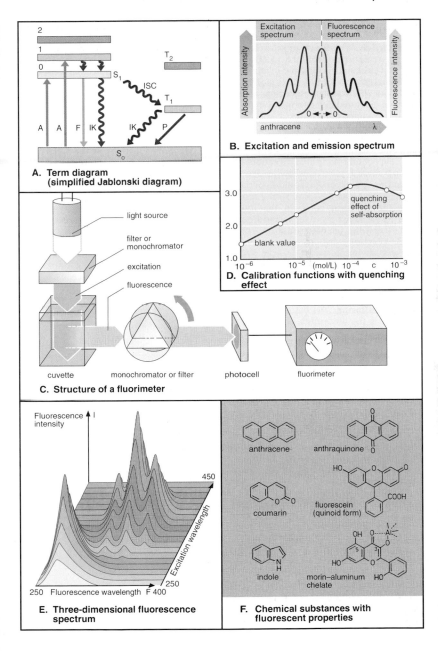

A. Term diagram (simplified Jablonski diagram)

B. Excitation and emission spectrum

C. Structure of a fluorimeter

D. Calibration functions with quenching effect

E. Three-dimensional fluorescence spectrum

F. Chemical substances with fluorescent properties

8.5 Infrared and Raman Spectroscopy

Introduction. With the absorption of light energy in the range from 0.8 to 500 μm, different mechanical vibrations of atoms or functional groups are excited in a molecule. In Raman spectroscopy, molecular vibrations are measured in the form of emission (Raman scattered radiation).

A. Harmonic vibration. We have used the harmonic vibration of a spiral spring as a simple model for molecular vibration. On the ends there are two balls having masses m_A and m_B (corresponding to atoms or parts of molecules), and the spiral spring represents the chemical bond. In the figure showing the stretching of the spring around x_0, starting with t_0 one obtains a cosine curve for the vibration equation $x = x_0 \cos(2nvt)$. Hooke's law applies: $K = -kx$ ($K =$ stretching of the spring; $k =$ bending force constant; $x =$ extension).

B. Comparison of IR absorption and Raman scattering. The basis of IR spectroscopy is the absorption of the energy from photons (hv) for the excitation of molecular vibration. Photons of a monochromatic light ray can undergo elastic collisions with molecular regions: the unshifted Rayleigh scattering occurs with loss of energy. With non-elastic collisions (where energy is lost), molecular vibrations are induced and the scattered photons are recorded as weak emissions in longer wavelength spectral region (Stokes region). If photons absorb vibrational energy when they collide (from excited molecular states), the (rarer) high-energy anti-Stokes radiation is produced.

C. Natural vibration of a three-atom bent molecule. In polyatomic molecules, each individual atom can move in the three directions of space. To describe the movements, $3N$ ($N =$ number of atoms) spatial coordinates are required (corresponding to $3N$ degrees of freedom). However, three of the movements in the same direction in space with concurrent position change of the mass center of gravity do not lead to molecular vibration. Three others only result in rotation around the center of gravity. Hence, for the whole of the degrees of freedom, $Z = 3N - 6$, and for linear molecules it is only $Z = 3N - 5$. (There is no rotation about the molecular axis; consider CO_2, for example.) A triatomic, bent molecule produces three normal vibrations: one symmetrical and one asymmetric valence vibration and a bending vibration (example: H_2O).

D. Normal vibrations of the CH_2 fragment. The normal vibrations described in Section C occur in the methylene group. With a difference of $80\,cm^{-1}$, the very intense bands (such as those of cycloalkanes) of the symmetrical and asymmetric valence vibrations are readily distinguishable.

E. Bending vibrations of the CH_2 fragment in alkanes. Bending vibrations include shear oscillation (see D.3) and rotary motion of the entire fragment about the Cartesian coordinates. They are called rocking, wagging and twisting modes of motion.

F. IR spectrometer. The light source of an IR spectrometer consists of a Nernst rod (ceramic material) which produces IR radiation when heating electrically to approximately 1600 °C. The radiation is split into two light beams having the same intensity. A swing mirror is used to alternatingly switch the measuring beam and the reference beam to the monochromator behind the cuvette. A thermoelement serves as the receiver.

G. IR cuvettes. Transparent parts are manufactured from crystals (disks) of NaCl, KBr and LiF. In addition to liquid cuvettes (with filling vents), gas cuvettes having larger thicknesses are also available.

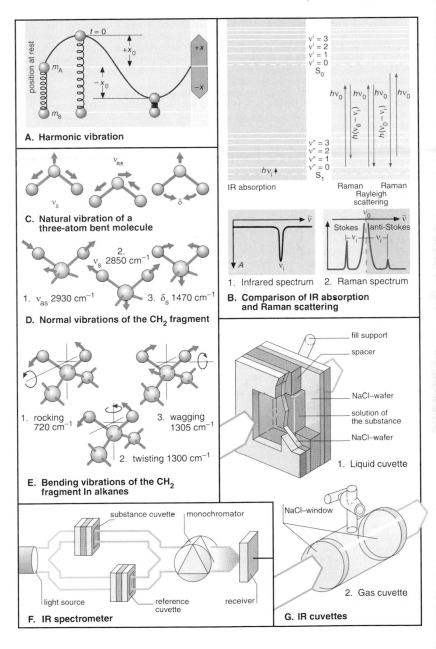

A. Harmonic vibration

C. Natural vibration of a three-atom bent molecule

D. Normal vibrations of the CH_2 fragment

1. ν_{as} 2930 cm^{-1}
2. ν_s 2850 cm^{-1}
3. δ_s 1470 cm^{-1}

E. Bending vibrations of the CH_2 fragment in alkanes

1. rocking 720 cm^{-1}
2. twisting 1300 cm^{-1}
3. wagging 1305 cm^{-1}

B. Comparison of IR absorption and Raman scattering

1. Infrared spectrum
2. Raman spectrum

F. IR spectrometer

G. IR cuvettes

1. Liquid cuvette
2. Gas cuvette

H. Valence and bending vibrations in the IR spectrum. Larger amounts of energy are necessary to induce valence vibrations than to induce bending vibrations. Above $v = 1500\,cm^{-1}$, predominantly valence vibrations occur, whereas binding vibrations predominate with lower wavenumbers (exception: NH bending vibrations above $1500\,cm^{-1}$). Valence vibrations of triple bonds and cumulative double bonds lie between 2800 and $2100\,cm^{-1}$ and those of double-bonded atoms between 2100 and $2500\,cm^{-1}$. If hydrogen atoms are involved, they lie between 4000 and $2800\,cm^{-1}$. Bands in the region from 1600 to $1000\,cm^{-1}$ can be traced back to backbone vibrations, and are characteristic of the entire molecule. Therefore, this region is called the fingerprint region.

The energy requirement for inducing a valence vibration is generally larger the greater is the bonding strength between the vibrating atoms and the smaller is the mass of these atoms. A high energy requirement for a vibration to be realized means a high wavenumber in the IR absorption spectrum. Accordingly, bands for double- and triple-bonded atoms should be found at higher wavenumbers than those for single-bonded atoms. However, since the hydrogen atom only has a small mass compared with C, N and O atoms, OH, NH and CH groups are found at high wavenumbers.

I. IR spectrum. IR absorption spectra are not recorded as a function of the wavelength, but rather of the wavenumber. The following information can be extracted from an IR spectrum: the examples for hexan-2-one $(CH_3CH_2CH_2CH_2COCH_3)$ (upper spectrum) and hexan-3-one $(CH_3CH_2CH_2COCH_2CH_3)$ (lower spectrum) show the group frequencies of CH [symmetrical stretching vibration (see C for stretching) at $3000\,cm^{-1}$] and of the CO (carbonyl) function (also stretching), labeled as CH and CO. The fingerprint region between 400 and $1400\,cm^{-1}$ (corresponding to 7 to 25 μm) is determined by the total molecule, i.e. by the interactions of all the atoms in the molecule. In this example, the group frequency can be established as band X in the upper spectrum as a characteristic band for a series of four or more methylene groups (for CH_2 chains). The fingerprint region can be used to differentiate between the two isomers in this case also.

In contrast to the UV–visible spectrum, in the IR spectrum the transmission T is usually recorded in the direction of increasing wavenumbers (an 'inverse' picture in comparison). If there is a high absorption, there is a large signal, which is negative with respect to the baseline. The location of the bands is often characterized by the wavenumber and the intensity is often characterized by the addition of s (strong), m (medium) or w (weak).

J. IR transmission of the solvent trichloromethane (chloroform). The following techniques come into question for recording an IR spectrum, for which 1 to 15 mg of material are required, depending on the method: measuring liquids (or molten substances) in the form of a thin film between two NACl disks, or measuring solid substances as suspensions in paraffin oil (Nujol) or in a mixture with KBr or KCl in the form of a pressed article. If solvents are used, they are liquids with few bands, such as carbon tetrachloride, trichloromethane or carbon disulfide, the natural spectrum of which must also be considered.

Fluorinated hydrocarbons such as trichlorotrifluoroethane (no CH bands) are used for the spectrometric determination of oils in water or contaminated soils, as per DIN regulations. The measurements are taken in the range between 3.4 and 3.5 μm (corresponding to a wavenumber around $3000\,cm^{-1}$).

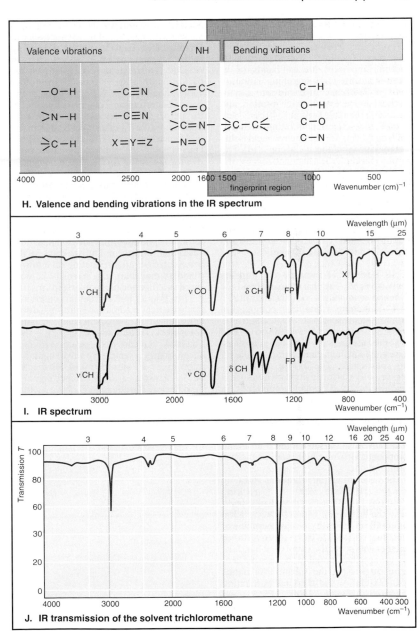

H. Valence and bending vibrations in the IR spectrum

I. IR spectrum

J. IR transmission of the solvent trichloromethane

K. Typical spectrum of saturated n-alkanes. The n-alkanes are relatively simple structures chemically, and therefore they present simple, easy to survey IR spectra. The most intense bands of n-alkanes are the C–H stretching vibration bands at $2890 \, cm^{-1}$ (a), split into a symmetrical and an asymmetric vibration, and the C–C–H bending band at $1340 \, cm^{-1}$ (c), as well as a scissoring vibrational band at $1500 \, cm^{-1}$ (b). Another intense band is due to a rocking vibration at $720 \, cm^{-1}$ (d). The carbon backbone is characterized by the multi-band region (fingerprint) between 1350 and $750 \, cm^{-1}$. This region can also be used for substance identification. For example, as a rule, cyclic hydrocarbons have more intense bands in the fingerprint region which come from ring distortion. Additional bands appear between 1250 and $910 \, cm^{-1}$ with branched chains.

L. IR spectra of benzene (1) and its derivatives (2). The characteristic bands of aromatic compounds at $1600–85 \, cm^{-1}$ (C–C stretching vibrations; conjugation with n-systems such as C=O, C=N, C=C cause a cleavage), $1500–1430 \, cm^{-1}$ and $700 \, cm^{-1}$ are due to C–C modes of motion; these bands are impacted relatively little by ring substituents. The $700 \, cm^{-1}$ band has an especially high intensity with mono- and *meta*-disubstituted aromatic compounds (in the region between 710 and $675 \, cm^{-1}$). If the substituents are different in *ortho*- and *para*-disubstituted products of benzene, there is also a weak absorption in this area. Symmetrically trisubstituted aromatic compounds also have high band intensities in this region.

The C–H stretching vibrations show up between 3100 and $3000 \, cm^{-1}$; rocking vibrations in the six-membered ring-level are in the fingerprint region between 1300 and $1000 \, cm^{-1}$. Finally, at wavenumbers of 2000 to $1600 \, cm^{-1}$, there are overtones together with several combination bands of CH groups. It is possible to identify aromatic rings using the substitution bands ($910–600 \, cm^{-1}$) and the C=C stretching vibrations around 1600, 1515 and $1450 \, cm^{-1}$.

M. IR spectrometry. To evaluate a substance quantitatively, it is first necessary to establish the zero and baselines. According to the baseline procedure, which makes it possible to exclude the effects of possible background noise (e.g. from tails from other bands), the zero line is drawn through the point of intersection C of the baseline and of the perpendicular through the absorption maximum. The T_0 value is derived from the difference between 100% and the zero line. One obtains the absorption value from $A = \log I_0/I = \log T_0/T$.

N. Fourier transform spectrometer. The region from 10 to $400 \, cm^{-1}$ is hard to access by conventional IR spectrometers. It can be studied using Fourier transform spectrometers. Based on the Michelson interferometer principle, the radiation first hits a beam-splitting mirror. The two halves of the beam interfere after reflecting on a mirror (fixed vs movable). The intensity is measured as a function of the interspin crossing s. Then Fourier transformation is used as a mathematical method for splitting a curve into a sum of sine and cosine functions using computers. The interferograms thus obtained contain the entire radiation absorption of a sample by wavelength and intensity as a Fourier sum of all spectral lines—shown as a cosine-shaped signal in the detector. The advantages of this method are improved resolution and higher energy flow.

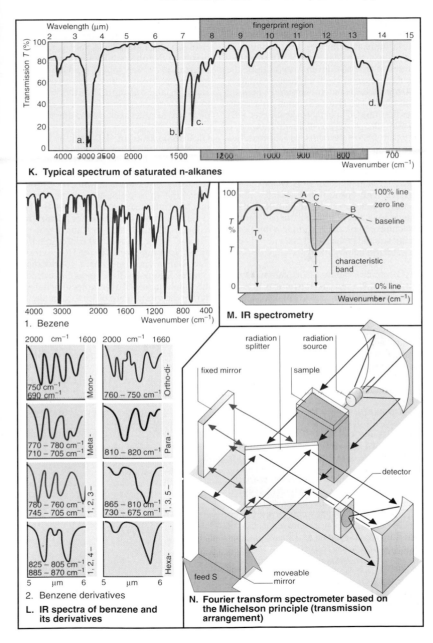

K. Typical spectrum of saturated n-alkanes

1. Benzene

M. IR spectrometry

2. Benzene derivatives

Mono-
750 cm⁻¹
690 cm⁻¹

Meta-
770 – 780 cm⁻¹
710 – 705 cm⁻¹

1, 2, 3 –
780 – 760 cm⁻¹
745 – 705 cm⁻¹

1, 2, 4 –
825 – 805 cm⁻¹
885 – 870 cm⁻¹

Ortho-di-
760 – 750 cm⁻¹

Para -
810 – 820 cm⁻¹

1, 3, 5 –
865 – 810 cm⁻¹
730 – 675 cm⁻¹

Hexa-

L. IR spectra of benzene and its derivatives

N. Fourier transform spectrometer based on the Michelson principle (transmission arrangement)

O. Examples of applications. *1. IR spectrum of a dried residue.* The IR spectrum of the dried residue from a water sample is shown as a practical example from environmental analysis. From 1 to 3 mg of the residue are ground well with approximately 100 mg of KBr and subsequently pressed with about 10^5 N in a vacuum in a compression mold. Under these conditions, KBr displays the properties of a cold flux, i.e. it traps the sample particles, and ultimately a transparent, tablet-shaped molded piece is produced.

A very characteristic, sharp and unambiguous band appears at $1390 \, cm^{-1}$ for nitrates. The absorption bands for phosphate, sulfates and silicates are less characteristic. Because of their similar structures, they have similar vibrational frequencies. The band at $1600 \, cm^{-1}$ is due to water molecules.

2. Correlation tables. Correlation tables are used to interpret an IR spectrum. In these tables the regions are already marked where absorption bands of a particular atomic grouping can occur. Correlation tables of the suspected substances or substance groups are consulted for the differential evaluation of the spectrum; such tables can be taken from the extensive collections of spectra currently available. Today, computers play an important role here also.

P. Raman spectrometer. A Raman spectrometer differs from an IR spectrometer (see F) particularly in the light source, which must provide intense monochromatic energy radiation. Today, lasers are usually used for this, as their beam can be focused on a sample. Plasma lines from the radiation of a laser (e.g. an argon or krypton laser) are filtered out by a pre-monochromator. The samples can be used as a pure liquid or solution, as a powered solid in a melting-point tube or as a gas in special multiple reflection cuvettes. A special cooling system is required for thermally or photolytically sensitive substances, or measurements are made while the sample is rotating. As a rule, the scattered light is only observed below an angle of 90°. Because of the very slight intensity of the Raman scattering radiation, the monochromator must be equipped with very high grade optics. 'Holographic' grids have very good resolution. In the photomultiplier, a photocell, the photons which appear produce a 'photoelectric current,' which is increased in the amplifier and transmitted further in signal form as a measurement for the intensity of the Raman scatter radiation to the recorder.

Q. Comparison of IR and a Raman spectrum. Dichloromethane (CH_2Cl_2) is used as an example to show the characteristic differences between an IR and a Raman spectrum. In the IR spectrum, the absorption of radiation is measured; in the Raman spectrum it is the intensity (see also B). The C–halogen bands suitable for identifying halogen alkanes are in the region of $726 \, cm^{-1}$. See J and K for the interpretation of the remaining bands in the IR spectrum. The Raman spectrum has a considerably simpler design. It is fundamentally true that the dipole moment of a molecule changes during the vibration in order to generate IR absorption. On the other hand, to obtain a Raman line of scatter, the polarizability must change during the vibration. Therefore, especially for symmetrical molecules, bands can occur in a Raman spectrum which are absence in the IR spectrum (and vice versa). Because of the differing physical principles, the intensities of the bands exist in a totally different relationship to one another. Raman spectroscopy can be used to identify nonpolar atom groups, to investigate symmetrical atom groups or molecules that are IR inactive and to perform quantitative analyses. However, structure analysis stands at the forefront of the applications.

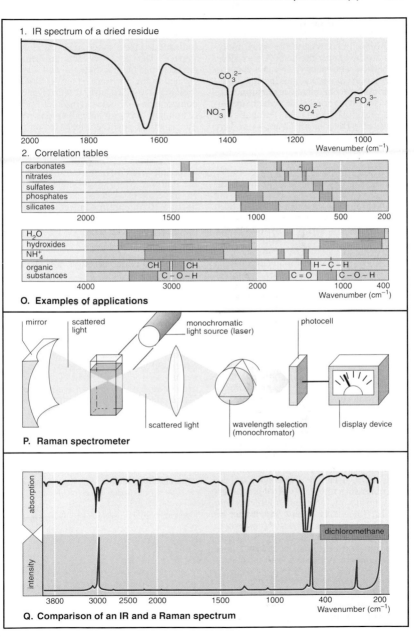

1. IR spectrum of a dried residue

CO_3^{2-}

NO_3^-

SO_4^{2-}

PO_4^{3-}

2000   1800   1600   1400   1200   1000
Wavenumber (cm^{-1})

2. Correlation tables

carbonates				
nitrates				
sulfates				
phosphates				
silicates				

2000   1500   1000   500   200

H$_2$O			
hydroxides			
NH$_4^+$			
organic substances	CH CH		H – C – H
	C – O – H	C = O	C – O – H

4000   3000   2000   1000   400
Wavenumber (cm^{-1})

O. Examples of applications

mirror | scattered light | monochromatic light source (laser) | photocell

scattered light | wavelength selection (monochromator) | display device

P. Raman spectrometer

absorption

intensity

dichloromethane

3800   3000   2500   2000   1500   1000   400   200
Wavenumber (cm^{-1})

Q. Comparison of an IR and a Raman spectrum

8.6 Mass Spectrometry

Definition. In contrast to the actual molecular spectroscopic methods (UV/visible, IR, NMR spectroscopy), in mass spectrometry it is the molecular ions (or fragments) resulting from chemical degradation reactions (following ionization) which are utilized for clarifying the structure of substances. The ions which are formed are separated in an analyzer (magnetic or electric field) based on their mass-to-charge (m/z) ratio (mass focusing).

A. Electron impact ionization and fragmentation diagram. The basic steps are shown for classical mass spectrometry from electron impact ionization and magnetic focusing: the ionization of molecules in a gaseous state with electrons (with an energy level of 10 to 15 eV; 1 eV = 96.5 kJ/mol) leads to the formation of molecular ions in that an electron is expelled from the molecule (with heteroatoms such as oxygen, sulfur and nitrogen, it is the free electron pairs). However, in actual practice higher energy levels of approximately 70 eV are used so that some of it will remain in the molecule after the ionization. If this excess energy corresponds at least to the activation energy of a decomposition reaction, there is a fragmentation, or a splitting of chemical bonds. In the first fragmentation step, a neutral molecule or a radical can be split off from the molecular ion, and a radical cation or a cation is formed. The latter can be split again in a second fragmentation step. When a radical ion decays, both types of fragmentation are possible; as a rule, for reasons of energy only a neutral molecule can be split off of a cation. Depending on the molecular structure, this is how characteristic, positively charged fragments are formed, since only certain bonds can be split. Molecular ions with little or no excess energy do not fragment: in the mass spectrum they form the 'mol peak,' and their mass number corresponds to the relative molecular mass.

B. Make-up of a mass spectrometer. A mass spectrometer consists of the following main components: in the inlet system, the sample is vaporized and is brought in a vapor state into the ion source, where the ionization takes place due to electron impact. The analyzer acts to separate (focus) the radical cations and cations formed in the ion source (see A) based on their mass-to-charge (m/z) ratio. The other components—detector, amplifier, printer or computer system—are all used to record the separated positive ions based on mass number (m/z) and frequency in the form of a mass spectrum. High-vacuum pumps generate pressures between 10^{-6} and 10^{-8} Pa in the inlet system, in the ion source and in the analyzer. Mass spectrometers are used as stand-alone systems or coupled (see Section 10.3), e.g. in connection with gas and liquid chromatography, or a plasma torch (such as inductively coupled plasma, ICP), as here for elemental analysis.

C. Cross-section through an electron impulse ion source. The vaporized substance flows from the inlet system to the ion source, where it is bombarded with electrons emitted from an electrically heated metal wire (filament). Between the filament and the target lies the chamber potential, which accelerates the electrons to the desired energy level. The filament and target are comparable to the cathode (e.g. incandescent cathode of an electron tube) and anode. The ion current is generated by the fact that the cations are pulled out of the region of ionization and fragmentation (the area of bombardment) and they are concurrently accelerated. The necessary acceleration potential lies between several hundred and several thousand volts, whereby the second electrode must be more negative than the first.

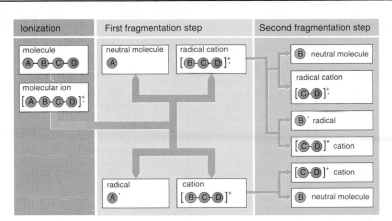

A. Electron impact ionization and fragmentation diagram

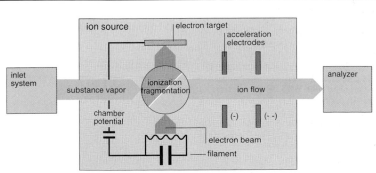

B. Make-up of a mass spectrometer

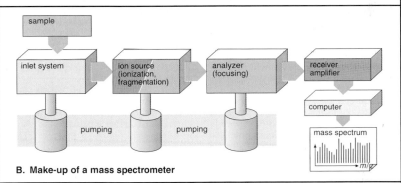

C. Cross-section through an electron impulse ion source

D. Monochromator portion of a mass spectrometer. During the flight of the positively charged particles through the magnetic field, the magnetic force and the centrifugal force maintain the equilibrium. The following equation applies: $m/z = \mu_r^2 \mu_0^2 r^2 / 2U$ ($m =$ mass of the particle; $z =$ its charge; $r =$ radius of the orbit; $\mu_r =$ permeability number; $\mu_0 =$ magnetic field constant; $U =$ acceleration potential). Thus the magnetic field has a dispersive effect on ions having different masses. This effect on different ion masses is comparable to that of a prism on polychromatic light.

E. The principle of magnetic focusing. After the cations formed in the ion source with the aid of the accelerator electrode arrive as an ion flow in the analyzer, a separation takes place here. This can be accomplished using a magnetic sector field in which the ions are brought into an angle of deflection of 60, 90, 120 or 180°.

The magnet analyzer in the figure consists of a metal tube, bent at an obtuse angle, and positioned between the poles of an electromagnet. The lines of field run perpendicular to the direction of travel of the ions (the ion flow).

The Lorentz energy K_L is decisive for the deflection of a positively charged fragment ion: an ion with mass m_1 attains a speed v_1 between the accelerator electrodes, and now it is subjected to the Lorentz energy in the analyzer at right-angles to its direction of acceleration. It produces flight paths with varying degrees of arch, depending on its mass (m_1, m_2). At a certain magnetic field strength (order of magnitude $1 T = 10000 G$), which can be altered, the flight path (of the particle with mass m_1 in this case) corresponds to the preset curvature of the analyzer tube.

The particle then hits the target and is recorded. Particles with larger or smaller masses impact on the walls of the analyzer tube. Thus, by continually increasing the magnetic field strength H, all cations with increasing mass will reach the target. This process is called magnetic focusing.

F. Directional focusing. By changing the magnetic field strength or the acceleration potential, ions having different masses can be brought into the same orbit as described. In practice, it is not possible to shoot into the magnetic field in only a single direction.

Therefore, in addition to its mass-dispersive property, the directional focusing effect of the magnetic field plays an important role. Ions of the same mass which enter the magnetic field in different directions are united again in a single location after the deflection.

G. Methods of ion separation. In addition to magnetic focusing, electrostatic and quadrupole focusing are frequently utilized. Electrostatic focusing (1) uses two condenser plates. Ions having the same energy are focused at the same location at the outlet slot, independent of the m/z value.

The principle behind quadrupole focusing (2) lies in a mass separation by vibrations of the ions in a high-frequency electric quadrupole field. The ion beam is directed in the longitudinal direction between four parallel metal rods; at opposite rods there is a d.c. voltage, the phases of which are shifted 180° and which are superimposed by a high-frequency field.

In a predefined current, which can be changed, only particles of a particular mass will reach the outlet slot along the path in the longitudinal direction between the rods.

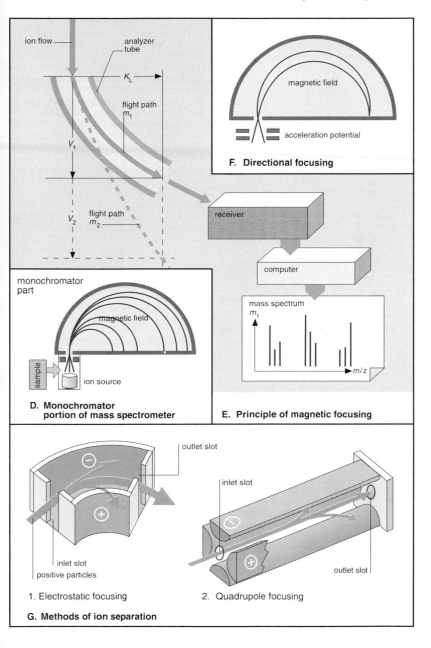

ion flow

analyzer tube

K_L

flight path m_1

V_1

V_2 flight path m_2

magnetic field

acceleration potential

F. Directional focusing

receiver

computer

mass spectrum m_1

m/z

monochromator part

magnetic field

sample   ion source

D. Monochromator portion of mass spectrometer

E. Principle of magnetic focusing

outlet slot

inlet slot

⊖

⊕

inlet slot
positive particles

1. Electrostatic focusing

inlet slot

⊖

⊕

outlet slot

2. Quadrupole focusing

G. Methods of ion separation

H. Double-focusing mass spectrometer.
Two signals are considered to be separated
if they overlap in 10% or less of their height.
Resolving power is defined as the ratio of
the mass number and the difference
between it and the mass number yet to be
resolved: $A = m/\Delta m$. For example, with
$m = 900$ and $\Delta m = 1$, $A = 900$. Most devices can achieve resolutions of approximately 2000. In a double-focusing mass
spectrometer, an electrostatic analyzer and
a magnetic analyzer are placed behind one
another. This method makes it possible to
achieve a resolution of up to 150000. With
this method, mass numbers can be determined experimentally, accurate to five
decimal places, so exact empirical formulas
of compounds can also be calculated
therefrom. From the outlet slot of the ion
source (a), the positively charged ions pass
through the electrostatic (b) and then the
magnetic analyzer (c), then after passing
through the collector slit (d) they meet at the
target collector (e). There are additional
collector slits in the zero-field ares between
(a) and (b) and between (b) and (c).

**I. The principle behind a time-of-flight
mass spectrometer.** Quadrupole (see G)
and time-of-flight mass spectrometers
contain dynamic ion-separation systems. In
a time-of-flight mass spectrometer, ions
having different mass are separated owing
to their differing times of flight for a prescribed pathway. The accelerated ions enter
the drift tube, in which lighter ions reach the
end faster than heavier ions. They are used
in atomic probes for the very sensitive
detection of short-lived particles, for
example.

J. Secondary ion microprobes. Mass
spectrometry is also used increasingly for
the characterization of the elemental make-
up of solid bodies: for the analysis of depth
distribution, of surfaces or in the form of
microregion analysis for two-dimensional
element distribution. If ions (secondary
ions) are formed by bombardment with
other ions (primary ions), this is called
secondary ion mass spectrometry (SIMS).
The ions which come from a primary ion
source pass through a monochromator and

ion optics before they impact on the analytical sample. The secondary ions thus
formed are analyzed with the aid of a double-focusing mass spectrometer. The latter
consists of both an electric and a magnetic
field with a monochromator interposed.

**K. Comparison of mass spectra follow-
ing different ionization methods.** In addition to electron impact ionization (EI), other
ionization methods we should mention are
the field ionization and field desorption
ionization methods. In field ionization,
positive ions are formed by the removal of a
strong electric field (on the anode). There is
little fragmentation due to the molecular
ion's weak energy.

To ionize compounds which are difficult
to vaporize (field desorption ionization), the
solution is applied to an activating metal
wire which is switched to the anode. After
applying the electric field, the ions formed
as desorbed.

Chemical ionization is one of the mildest
ionization methods: the sample is introduced into the usual electron impact ion
source along with methane at approximately 10^{-8} Pa. Methane cations are
formed which can transport protons to non-
ionized methane molecules due to the
relatively high methane pressure (10^2 Pa).
These CH_5^+ cations (isobutane cations in
this case) ionize the substance molecules by
giving up protons.

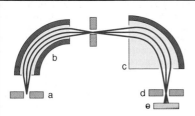

H. Double-focusing mass spectrometer

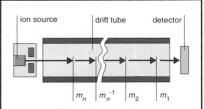

I. Principle behind a time-of-flight mass spectrometer

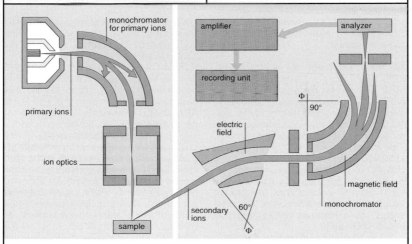

J. Secondary ion microprobes

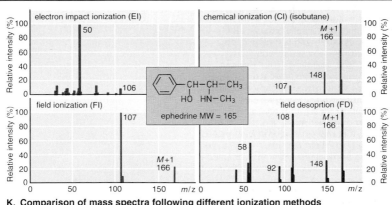

K. Comparison of mass spectra following different ionization methods

L. Comparison of field and electron impact ionization. With a molecular weight of 150, the molecular peak for ribose cannot be picked out of the mass spectrum after electron impact ionization owing to its fragmentation.

The gentler method of field ionization, in which lower energy molecular ions are formed, displays this signal as well as clear differences in the fragmentation. The appearance of the (low-energy) molecular ion M^+ and of the cation $M + 1$ (m/z 151) is the basic characteristic of this method, which can also be executed especially advantageously in the form of field desorption ionization.

M. Mass spectrum of n-tetradecane. Mass spectroscopic decomposition reactions are monomolecular, endothermic processes. Relative bonding strengths, steric factors and the stability of the products formed in the ion source influence the probability of the decomposition reaction. The energy capacity of the ion formed (depending on the stability of the positive charge) is particularly decisive for the course of the fragmentation reaction. In a high vacuum, isolated ions are present which undergo a conversion of electrical energy into vibrational energy, which is used to split bonds.

When alkanes are ionized, an electron is knocked out of a σ-bond. Owing to the resultant single electron bond, this bond loosens up, which in turn provides favorable conditions for bond breakage. With the exception of the terminal groups, all C—C σ-bonds are split with a similar degree of probability. An ion series with $C_nH_{2n+1}^+$ is formed. The fragments pass through an intensity maximum at C_3 to C_4; the intensity decreases with increasing carbon number. Secondary cleavage (such as the splitting off of olefins) leads to the favoring of lighter fragments in the mass spectrum of n-alkanes. If branches develop in an aliphatic hydrocarbon, then the ions formed by cleavage at these site show greater intensities than their homogeneous neighbors.

N. Fragmentation and mass spectrum of n-butylbenzene. In general, intensive signals of molecular ions appear (2) in the mass spectra of aromatic hydrocarbons. Compared with the aliphatic hydrocarbons, fragment ions appear less frequently (1). Doubly charged ions are recorded with increasing frequency.

The most important fragmentation reactions with alkylbenzenes such as n-butylbenzene are as follows. Cleavage of the benzylic activated bond (a) leads to the formation of benzyl cations (b), which are in an equilibrium with the isomeric tropylium cation (c). This ion, with m/z 91, is especially stable and appears as a characteristic for a simple benzyl cleavage as a signal with high intensity. By splitting off ethyne, the pentacyclodienyl cation, with m/z 65, is formed. Cleavage of bonds further removed from the benzene ring yields fragments with decreasing intensity. Homologous ions with mass $91 + 14n$ ($n = 1, 2, 3, \ldots$) are formed. Olefin molecules are formed from the molecular ion of alkylbenzenes with a side chain of at least three C atoms.

The formation of the olefin molecule (d) (radical cation) also increases as the size-chain length increases; it competes with the benzyl cleavage just described. Other aromatic compounds which cannot fragment according to these principles split off H_2 and C_2H_2 (ethyne). Doubly charged cations are formed as a result of ionization (as are observed with anthracene, for example).

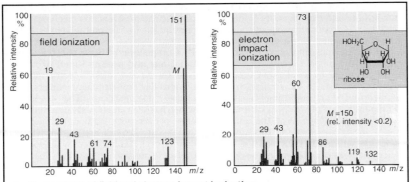

L. Comparison of field and electron impact ionization

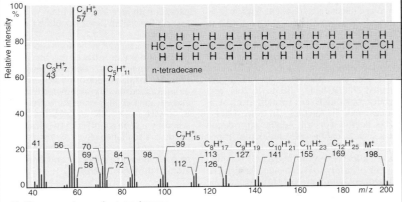

M. Mass spectrum of n-tetradecane

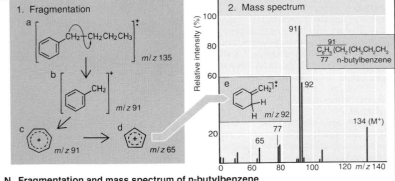

N. Fragmentation and mass spectrum of n-butylbenzene

O. Comparison of mass spectra of cyclohexane and cyclohexene. Cycloalkanes basically produce mass spectra similar to those of alkenes. However, since rearrangement reactions can also occur, they are less clear. The simple cleavage of a bond, which can occur by an electron breaking out, does not lead to the formation of a fragment ion. After the splitting of a σ-bond and the formation of a radical cation, this initiates an α-cleavage. With the elimination of an olefin—ethene in this case—a $C_4H_8^{+\cdot}$ ion with m/z 56 is formed.

In cyclohexene, ionization of the double bond also induced an α-cleavage and therefore an opening of the ring system. As with cyclohexane, no fragments are produced at this point. Only after a second α-cleavage occurs is a fragment with m/z 54 formed. This ion is formed as a result of retro-Diels–Alder cleavage (RDA).

Diels–Alder cycloaddition (diene synthesis) involves the addition of a $C=C$ double bond, activated by neighboring polar groups, to hydrocarbons with a conjugated double-bond system. In this way cyclic compounds are formed from 1,3-dienes and dienophiles. In retro-Diels–Alder degradation, a 1,3-diene is produced as a radical cation [M = molecule minus C_2H_4: $(M-C_2H_4)^{+\cdot}$], ethene (C_2H_4) is the dienophile. The ion with m/z 81 is formed as a result of an allylic H cleavage. Finally, the subsequent elimination of H_2 produces the particularly stable phenomium ion with m/z 79; an additional H_2 cleavage explains the presence of the signal at m/z 77.

Another important reaction is the elimination of a methyl radical ($CH_3 \cdot$): after the allyl cleavage, there is the rearrangement of a hydrogen atom, so that a primary radical is converted into an allylic radical. A second allyl cleavage yields the most intense ion $(M-CH_3)^+$ in the mass spectrum with m/z 67. Thus, all degradation products are induced by the location of the radical in the molecule.

On the other hand, in the retro-Diels–Alder cleavage the splitting off is accomplished via the elimination of an especially stable neutral particle (olefin).

It can be seen that a knowledge of basic organic chemical reaction mechanisms is a prerequisite for interpreting mass spectra.

P. Comparison of mass spectra of pentan-2-one and pentan-3-one. A comparison of the two mass spectra makes it obvious that even structural isomers can be clearly distinguished. The α-cleavage which has been mentioned numerous times also occurs with carbonyl compounds (aldehydes, ketones and carboxylic acid derivatives). It is triggered by the localization of a radical on oxygen. The result mesomerically stabilized acylium ions eliminate carbon monoxide in a secondary reaction. The most intense signals are obtained via these mesomerically stabilized cations with m/z 57 for pentan-3-one and m/z 43 for pentan-2-one according to their structures.

The ion at m/z 58 in the mass spectrum of pentan-2-one can be traced to a McLafferty rearrangement. This is also a rearrangement with the displacement of hydrogen. An H atom is added via a six-membered transition state to another atom or an atom group (e.g. S, —CH_2, or =NR in addition to O), whereby a splitting off into two fragments occurs simultaneously. A McLafferty rearrangement occurs any time there is a multiple bond in a molecule and there is a H atom in the γ-position to it.

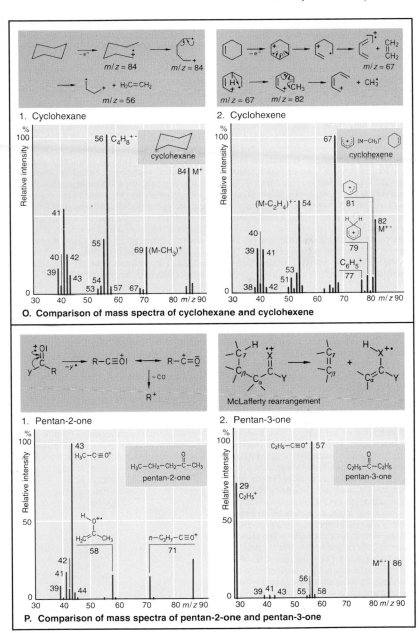

O. Comparison of mass spectra of cyclohexane and cyclohexene

1. Cyclohexane

$m/z = 84$
$m/z = 84$
$m/z = 56$
$+ H_2C=CH_2$

%
100
56 $C_4H_8^{+\cdot}$
cyclohexane
84 M^+
41
40 42
39 43
55
54
53 57 67
69 $(M-CH_3)^+$
30 40 50 60 70 80 m/z 90

2. Cyclohexene

$m/z = 67$
$m/z = 67$ $m/z = 82$ $+ CH_3^{\cdot}$

%
100
67 $(M-CH_3)^+$ cyclohexene
$(M-C_2H_4)^{+\cdot}$ 54
81
82 $M^{+\cdot}$
40
39 41
79
53
51 $C_6H_5^+$
38 42 77
30 40 50 60 70 80 m/z 90

P. Comparison of mass spectra of pentan-2-one and pentan-3-one

$R-C\equiv\overset{+}{O}| \longleftrightarrow R-\overset{+}{C}=\overline{\underline{O}}$
$-y^\cdot$
$\downarrow -CO$
R^+

McLafferty rearrangement

1. Pentan-2-one

%
100
43 $H_3C-C\equiv O^+$
$H_3C-CH_2-CH_2-\overset{O}{\overset{\|}{C}}-CH_3$
pentan-2-one
$H_2C \overset{H}{\underset{\|}{\overset{\overset{+\cdot}{O}}{C}}} CH_3$
58
$n-C_3H_7-C\equiv O^+$
71
42
41
39 44
30 40 50 60 70 80 m/z 90

2. Pentan-3-one

%
100
$C_2H_5-C\equiv O^+$ 57
$C_2H_5-\overset{O}{\overset{\|}{C}}-C_2H_5$
pentan-3-one
29 $C_2H_5^+$
50
56
39 41 43 55 58
$M^{+\cdot}$ 86
30 40 50 60 70 80 m/z 90

8.7 Nuclear Magnetic Resonance (NMR) Spectroscopy

A. Atomic nuclei in a magnetic field.

Atoms with nuclei containing an odd number of either protons or neutrons have a magnetic moment induced by the nuclear spin. The nuclei ^1H, ^{13}C, ^{14}N, ^{19}F and ^{31}P have a spin quantum number $I = 1/2$ and therefore they have a magnetic moment, meaning they can be used to perform NMR spectroscopy. A spin quantum number $1/2$ means that their magnetic moment μ can only have two equal but opposite values, $+\mu$ and $-\mu$, with $m = +1/2$ and $-1/2$. Both settings of the magnetic moment μ in a magnetic field B_0 have only slight differences in energy. m is the magnetic quantum number with $m = I, I - 1, I - 2, \ldots$. The nuclear moments can be oriented parallel $(m = +1/2)$ or anti-parallel $(m = -1/2)$ to the field; the parallel orientation is favored energetically.

B. Spin of the hydrogen nucleus.

The moving, positively charged proton forms a magnetic field in a metallic conductor, just like electrons in an electric current. The proton behaves like a bar magnet with a north and a south pole. The direction of nuclear rotation (spin direction) also determines the direction of the magnetic field (indicated by arrows).

C. Nuclear resonance and relaxation.

We will use a mechanical top as a model for the behavior of a hydrogen nucleus in a magnetic field. As a top rotates about its x-axis, if something impacts it, it also begins a staggering movement, called precession. In an external magnetic field (with field strength H_0 and flow density B_0), the hydrogen nucleus rotating about its axis ('nuclear top') also experiences an impact of sorts: the resulting movement proceeds in two preferred directions (directional quantization), associated with an energy division (or quantization): (1) in the direction of the external field (with E_1, lower energy and therefore more likely) and (2) in the opposite direction (E_2). The 'impact' is generated using a high-frequency transmitter; it causes an inversion of the magnetic vector.

The energy (δE) required to convert the nucleus from the lower energy state (E_1) to the higher energy state (E_2) is called the resonance energy. Nuclei with higher energy (E_2) return to the lower energy position after a short time; this process is called relaxation.

D. Chemical shift.

NMR spectroscopy registers the energy needed to convert to the next energy level as a signal of the associated frequency. The position of these signals, i.e. the amount of energy required for nuclear resonance, called the chemical shift, depends on the binding ratios of the particular atom. The resonance positions can be generally characterized as follows: (1) with a hypothetical atom without an electron shell, (3) with a real atom, (2) with electron-attracting neighboring atoms and (4) with electron-pushing neighboring atoms. The frequency of the resonance is proportional to the applied magnetic field.

E. Inductive effect.

If an atom is brought into a homogeneous magnetic field, the movement of the electrons surrounding the nucleus in the nuclear area induces a secondary field, opposite to the applied field. This produces a magnetic shielding of the nucleus, which can be reduced as a result of the inductive effect of neighboring electron-negative atoms. Therefore, protons of a methyl group bound to an oxygen (CH_3—O—) absorb as protons of an N-methyl group (CH_3—N=) in a deeper field. The chemical shift is indicated in ppm (δ as the distance of a signal from that of a standard compound, divided by the resonance frequency), with band gaps of several hundred Hz and resonance frequencies of MHz; therefore, ppm (parts per million).

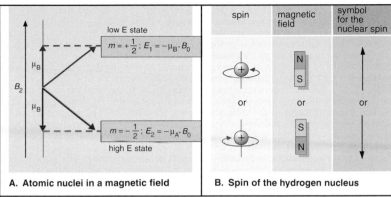

A. Atomic nuclei in a magnetic field

B. Spin of the hydrogen nucleus

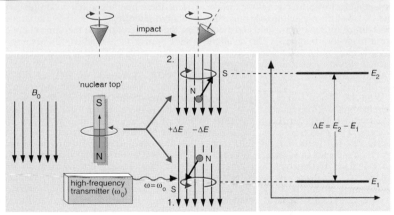

C. Nuclear resonance and relaxation

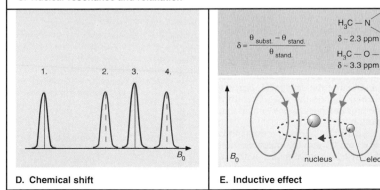

D. Chemical shift

E. Inductive effect

F. Anisotropy effect. In addition to the properties of neighboring groups, the chemical shift in the NMR spectrum of a proton depends on the spatial arrangement—on the distance and the bond angle to the proton (see the example of a carbonyl group in 1). The chemical shift is caused especially by anisotropy of π-bonds. In general, anisotropy refers to the different effects of physical and chemical forces in the different directions of space. The anisotropy of the carbonyl group (2) causes an increasing shielding of protons in the regions marked with a $+$ sign versus those marked with a $-$ sign. In contrast to inductive effects, anisotropic effects are field effects which affect the surrounding space. If the molecule is perpendicular to the direction of the lines of field (B_0), then the π-electrons can move in a circle, inducing a magnetic field and therefore a magnetic dipole (3). The resultant strengthening of the outer field at the location of the proton leads to a shift of the bands to a higher frequency.

G. Van der Waals forces. If protons are brought together close enough for a van der Waals repulsion to occur, then the shielding weakens. The proton, in a cyclohexanone system (1) H* (with a rigid chair conformation) appears in a lower field in resonance for $R = CH_3$ than for $R = H$. In (2) the molecule (1) is shown with a view toward the carbonyl group: the steric interaction between the groups H and R exclusively causes a weakening of the shielding.

H. Anisotropy effects with aromatic compounds. If a benzene molecule is positioned perpendicular to the lines of field, the π-electrons start to rotate (see F) here also. In so doing, they generate a dipole counter to the external field. The external field is weakened owing to the lines of force above and below the ring level, but it is strengthened outside the ring. Therefore, aromatic protons appear at lower field than olefin protons.

For resonance to occur, the transmitter frequency (see C) and the magnetic field strength must be synchronized with one another: the field which is effective at the nucleus (B_{eff}) is derived from $B_0(1 - \sigma)$, where σ is the shielding constant.

I. Spin–spin coupling. Fine structures in an NMR spectrum can be traced to interactions of the nuclei with neighboring magnetic atoms, or spin–spin coupling. This can be transferred via space and via binding electrons. By definition, the coupling constant J has a plus sign if the state is stabilized with an anti-parallel orientation of the nuclear spin due to the coupling. Spin–spin coupling can occur between H and F, or between atoms of the same isotope with different chemical shifts (1). The example of fluoromethane (CH_3F) shows the splitting of the fluorine signal by the same three neighbors in a quadruplet with relative intensities $1:3:3:1$. The protons of the CH_3 group with the same chemical shift appear as a doublet (2) because of coupling with the fluorine atom.

J. NMR spectrum of HF. In the HF (hydrogen fluoride) molecule, the magnetic moments of both nuclei are oriented with respect to the magnetic field B_0: a weakening of the magnetic field at the fluorine nucleus occurs if the magnetic moment of the proton points in the direction of the field. If it points in the direction opposite to the field, there is a slight increase in field strength. The energy levels (see A) are split owing to the coupling of the fluorine atom with the proton; therefore, the fluorine resonance appears as a doublet (2) in the NMR spectrum. Spin–spin coupling is independent of the field strength.

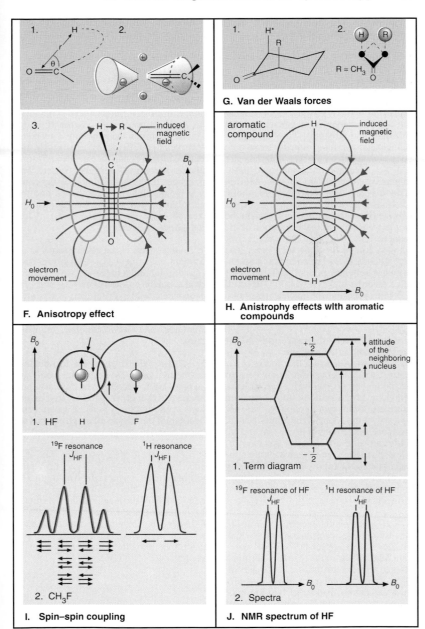

F. Anisotropy effect

G. Van der Waals forces

H. Anistrophy effects wIth aromatic compounds

I. Spin–spin coupling

1. HF

2. CH$_3$F

^{19}F resonance J_{HF}

^{1}H resonance J_{HF}

J. NMR spectrum of HF

1. Term diagram

$+\frac{1}{2}$

$-\frac{1}{2}$

attitude of the neighboring nucleus

2. Spectra

^{19}F resonance of HF J_{HF}

^{1}H resonance of HF J_{HF}

K. Schematic diagram of an NMR spectrometer. To conduct NMR measurements, one requires a homogeneous magnetic field with field strength H_0, a high-frequency transmitter to irradiate the sample and a measuring device for recording the resonance energy absorbed as a factor of frequency or field strength. The central point of the device is a strong electromagnet with $1-2$ T (superconducting magnets have a 10-fold field strength). To obtain good homogeneity for sharp signals, large pole shoes and small measuring volumes are used. To determine horizontal field inhomogeneities, one lets the sample solution rotate in a glass tube. The electromagnetic radiation of the high-frequency transmitter generated by the transmitter coil acts on the sample perpendicular to the direction of the magnetic field H_0. The receiver coil is arranged around the glass tube here; it is perpendicular to the transmitter coil and to the magnetic field H_0. Coupling the receiver with a recorder makes it possible to record an NMR spectrum, the nuclear magnetic resonance signal as a factor of the magnetic field strength H or of the frequency v.

L. Energy level diagram and NMR spectrum of nitropropane. *1. Energy level diagram.* In propane, the protons in the terminal methyl groups are equivalent to one another; the six methyl protons split the methylene group protons into a seventh multiplet. Spin–spin coupling is not observed with more than three bonds. Therefore, the protons of groups (a) and (c) in nitropropane are split by the two (b) protons in triplets. The (b) protons are split twice: one by the (a) protons and again by the (c) protons. The energy level diagram for the methylene (b) protons with a total of 12 lines is shown here.

2. NMR spectrum. However, some of the lines in the energy level diagram are so weak that they do not appear in the NMR spectrum. The triplets for the (a) and (c) protons do not exactly exhibit the $1:2:1$ signal peak ratios. This phenomenon always occurs when the chemical shifts of two kinds of protons are very similar. Three kinds of information can be derived from an NMR

spectrum: the chemical shift, the spin–spin coupling pattern and the peak areas, which are proportional to the number of protons that generate a resonance signal.

M. ^{1}H resonance lines of ethyl groups. The general rules for the line intervals and for the relative intensity of the multiplets are valid with either different nuclei or nuclei of the same isotope with chemical shifts around 10 times the coupling constants. With smaller differences, changes in the relative intensities occur (see also L)—one obtains spectra of a higher order. The intensities of the lines lying near the common focal point are strengthened and the outer-lying ones are weakened (see the example with nitropropane in L). Additional lines also occur.

1. Diethyl ether. Using ethyl groups as an example, with $-O-CH_2-CH_3$ groups both a quadruplet and a triplet occur—symmetrical about the position of the CH_2 and CH_3 resonances, with a difference in the chemical shifts of approximately 2.2 ppm (198 Hz in the 90 MHz spectrum), simple structures and whole-number intensity ratios.

2. n-Butane 90 MHz spectrum. In n-butane, the methyl and ethyl resonances only lie 0.4 ppm apart. Therefore, in this case, one obtains a higher order spectrum from many lines: spin couplings and the exact band locations of the CH_2 and CH_3 groups can no longer be easily recognized or assigned.

3. n-Butane 400 MHz spectrum. The distance between two bands is generally proportional to the magnetic field strength. On the other hand, the coupling constants are independent of the magnetic field strength. By increasing the frequency from 90 to 400 MHz, one obtains a greater distance between the bands; also, the triplet and quadruplet structures are more easily recognized.

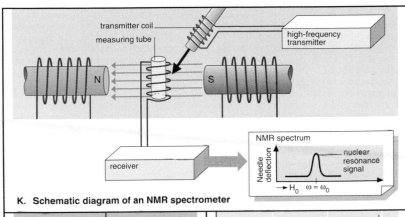

K. Schematic diagram of an NMR spectrometer

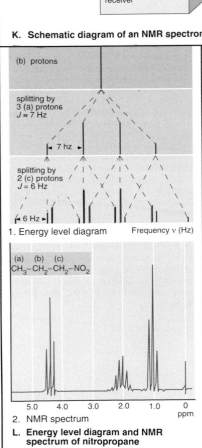

(b) protons

splitting by
3 (a) protons
$J \approx 7$ Hz

⊢ 7 hz ⊣

splitting by
2 (c) protons
$J = 6$ Hz

⊢ 6 Hz ⊣   Frequency ν (Hz)

1. Energy level diagram

(a)   (b)   (c)
$CH_3 - CH_2 - CH_2 - NO_2$

2. NMR spectrum

L. Energy level diagram and NMR spectrum of nitropropane

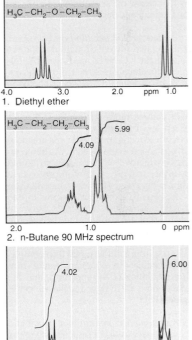

$H_3C - CH_2 - O - CH_2 - CH_3$

1. Diethyl ether

$H_3C - CH_2 - CH_2 - CH_3$

5.99

4.09

2. n-Butane 90 MHz spectrum

4.02

6.00

3. n-Butane 400 MHz spectrum

M. ^{1}H resonance lines of ethyl groups

N. ^{13}C-NMR spectrum of 2,2,4-trimethylpentane. Although ^{13}C-NMR spectroscopy was not developed until 10 years after ^{1}H-NMR spectroscopy (roughly in the mid-1950s), it has reached even greater significance in the meantime. Because of the low percentage of ^{13}C in the natural isotope mixture of carbon (1.1%) and because of the considerably smaller gyromagnetic constants (proportionality factor; refer to physics texts about a top's precession movement and Lamor equation) compared with that of the proton, the intensities of the ^{13}C and ^{1}H signals (at the same C and H concentrations) are in a ratio of 1 : 6000. To increase the sensitivity of ^{13}C-NMR measurements, the summation of numerous spectra is performed. Another way to increase the sensitivity is to use the proton broadband decoupling method.

1a. Without proton uncoupling. Dissolved in a deuterated solvent (dimethyl sulfate in this case), the ^{13}C-NMR spectrum of 2,2,4-trimethylpentane yields the following general characteristics: the range of chemical shifts is considerably larger than in ^{1}H-NMR spectra. Therefore, the spectra as a whole contain more information. Signals overlap less frequently; structural differences can be recognized better and are more easily assigned to the signals. Carbon couplings are relatively strong and also extend out further spatially than with protons. In addition, because of the low concentration of ^{13}C, two ^{13}C atoms rarely occur in one molecule. Therefore, couplings between carbon atoms are not found in the spectra. On the other hand, coupling with neighboring protons and with those further away leads to splitting of the signal into complicated multiplets. A widening results in a better resolution of the spectrum, and the signal regions are assigned to the molecule regions (1b).

2. Broadband decoupling. In a broadband decoupled spectrum, all resonances appear as singlets. Signal intensities are increased, and all ^{1}H/^{13}C couplings are removed. A broad frequency band is radiated into the sample, and the radiated electromagnetic vibration is modulated with low-frequency interference. Thus, the resonance frequency is conveyed to each coupled hydrogen nucleus. In so doing, it changes its spin orientation so rapidly that a magnetic field no longer appears and therefore there is no coupling.

3. Off-resonance decoupling. In this technique, the ^{1}H/^{13}C couplings are retained, but the coupling constants are reduced. The decoupling frequency is set outside the resonance area of the protons. The frequency is not in direct resonance with the decoupled nucleus, but it still affects the nucleus. Now one obtains quartets, triplets and doublets for CH_3, CH_2 and CH groups with slight overlap.

O. Proton off-resonance decoupling. Another decoupling technique, proton off-resonance decoupling, is compared with broadband decoupling using a pharmaceutical agent [2-(4-chlorophenoxy)-2-methylethyl propionate, used to reduce the risk of heart attack]. A decoupling frequency is applied outside the proton resonance region. Although it is not in direct resonance with the nucleus to be decoupled, it still affects the nucleus. In this way one obtains quartets (Qu), triplets (T) and doublets (D) for the CH_3, CH_2 and CH groups and singlets (S) for the quaternary carbon.

In contrast to the broadband decoupled spectrum, in this case the ^{1}H/^{13}C couplings are retained; their coupling constants are merely reduced. The splittings overlap less and are disturbed less by ^{1}H/^{13}C remote couplings (according to Rücker, Neugebauer and Willems, 1988; see Bibliography).

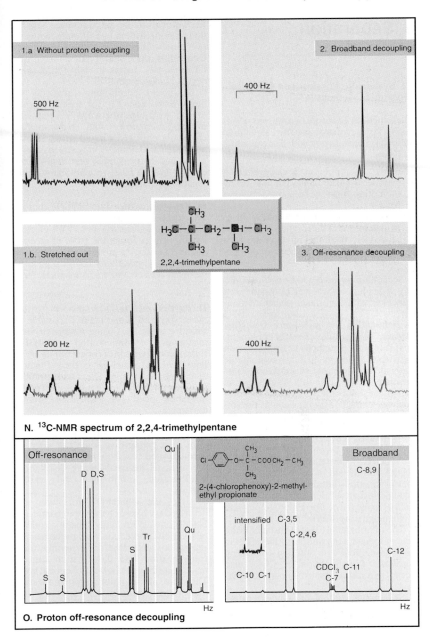

N. ^{13}C-NMR spectrum of 2,2,4-trimethylpentane

O. Proton off-resonance decoupling

9 Separation Methods

9.1 Systematic Physical and Chemical Separation Methods

Introduction. There are two reasons why it might be necessary to use separation methods in conjunction with determination methods:

1. If the selectivity of a determination method is insufficient, the interfering impurities must be separated beforehand.
2. If as many substances as possible are to be detected and determined in a single procedure, then the mixture must be separated into its respective components.

Separation methods can be divided initially depending on whether or not they are associated with a substance transformation.

A. Separation via substance transformation. This includes classical gravimetry as a precipitation method (1): the separation steps for cations in qualitative inorganic analysis are also based on substance transformation. The differing solubilities of individual compounds or groups of compounds—principally with water as the solvent—are used for the separation. In trace determinations of elements, coprecipitations are used to enrich low concentrations in voluminous precipitates. In electrogravimetry, substance transformation occurs with the aid of electric current (2): ions in aqueous solution are converted into neutral substances (elements or compounds such as oxides) (see also Polarography and Voltammetry, pp. 62*ff*).

The volatilization (3) of a substance in the aqueous phase, for example by the addition of a strong acid to a solution containing carbonate (volatilization of CO_2), also occurs as a result of substance transformation.

Separations without substance transformation can be accomplished based on partitioning procedures, different particle charges, particle masses, sizes and shapes and different vapor pressures.

B. Separation by partitioning. Separations due to different partitioning between two immiscible phases are very important. Depending on the phase types involved, one differentiates between: adsorption processes (solid–liquid or solid–gas) (1); liquid–liquid partitioning (generally as liquid extraction) (2); the special process of ion exchange (between a solid ion-exchange matrix with substitutable, charged groups and a liquid phase, usually water) (3); and liquid–solid extractions (4). As separation principles, partitioning procedures are the basis for chromatographic separation methods.

C. Separation based on particle charge. The different migration of free charged particles in an electric field with electrokinetic effects forms the basis of electrophoretic separation procedures (1). In mass spectrometry, positively charged, gaseous particles are separated in a magnetic field (2).

D. Particle effects. Differences in particle mass, size and shape make it possible to separate substances using molecular sedimentation (sedimentation analysis), (ultra)centrifugation (1) or gel filtration (2). Pore sizes and structures of the gel matrix determine the filtration effect (gel filtration, filtration through a molecular sieve) or the migration speed (gel permeation chromatography).

E. Vapor pressure differences. Separations on a preparative scale can often be accomplished based on the different vapor pressures of materials using distillation, condensation, sublimation and crystallization.

All of the separation procedures just described are based on thermodynamic and kinetic laws. With many separation methods, various procedures lead to the desired separation, such that no generally valid categorization and demarcation can be made according to physical-chemical laws. Therefore, all attempts at systematization must remain more or less arbitrary. For chromatography, the categorization (see p. 146) based on phase pairs and separation method forms the basis of any systematization.

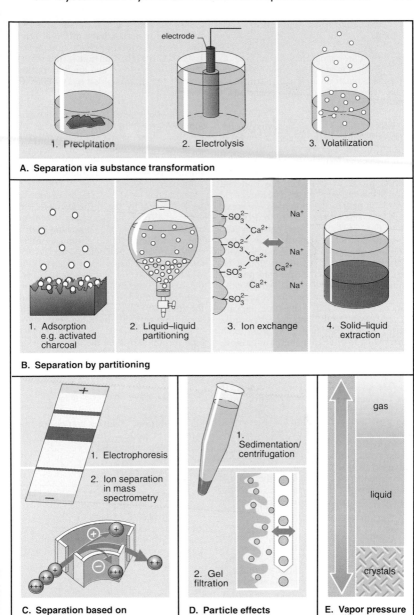

A. Separation via substance transformation

1. Precipitation
2. Electrolysis
3. Volatilization

electrode

B. Separation by partitioning

1. Adsorption e.g. activated charcoal
2. Liquid–liquid partitioning
3. Ion exchange
4. Solid–liquid extraction

C. Separation based on particle charge

1. Electrophoresis
2. Ion separation in mass spectrometry

D. Particle effects

1. Sedimentation/centrifugation
2. Gel filtration

E. Vapor pressure differences

gas

liquid

crystals

9.2 Chromatographic Separation Methods

The term chromatography is applied to those physical methods in which there is a substance transformation via partitioning between a resting (stationary) phase and a moving (mobile) phase.

A. Kinetic theory. *1. Chromatographic process.* Kinetic theory assigns a selective affinity for the stationary phase to individual molecules (circles, triangles and rectangles in the figure). It is the task of the mobile phase to transport the material through the stationary phase. The longer the separation path is, the more they are separated from one another owing to different intensities of interactions between the materials and the stationary phase. If one compares the individual substances with boats, the mobile phase with a river and the stationary phase with the river bank and the landing places along the bank, we have the following figurative model: the boats (substances, a) all start at the same time, they are moved forward by the river current (mobile phase, b), and they pull away from each other owing to different patterns of motion, stopping for differing lengths of time on the river bank (stationary phase, c and d).

2. Thin-layer chromatogram product. If one conducts this type of separation on a thin layer of an adsorbent such as silica gel, then after the mobile phase has migrated a distance c, the substances remain on the layer as spots A and B. The position of these spots is indicated as a retention factor (R_f) value as follows: $R_f(A) = a/c$, $R_f(B) = b/c$. The distance c is traveled by the mobile phase as a whole; distances a and b are the characteristic retentions for substances A and B.

B. Theoretical plate model. *1. Elution chromatogram.* If the stationary phase is packed into a column, the separated materials can be completely extracted (eluted) from the column using the mobile phase. This process is depicted in the form of a chromatogram, or elution curve, which shows the amount (or concentration) of the eluted substance as a factor of time. The individual materials have different retention times; the total retention time t_R consists of the net retention time t_S (retention in the stationary phase) and the flow-through time of the mobile phase t_m (without any retention).

2. Gaussian curve. In the ideal case, an elution curve will show a Gaussian distribution. The shape is determined by diffusion processes and by statistical irregularities in the establishment of equilibrium between the substance and the stationary phase. The broadening phenomenon (see C) as a factor of time can be explained by the diffusion. The substances diffuse further apart from one another the longer the separation path is. The distribution of a substance after a chromatographic procedure in the form of a curve is also called a band or a peak, the width of which (e.g. at half-height $b_{0.5}$ or at the base w) depends on the distance traveled. The theoretical plate model breaks the stationary phase of a chromatographic separation path into its individual sections or layers, the theoretical plates, in which an equilibrium between the stationary and mobile phase exists. The number of theoretical plates is a measurement of the efficiency of a separation column:

$$N = 16(t_R/w)^2 = 8 \ln 2(t_R/b_{0.5})^2$$

C. Resolution, relative retention and number of plates. The resolution of two bands is defined as follows:

$$R = 2(t_{R_2} - t_{R_1})/(w_1 + w_2)$$

The following statements can be made about the pairs of bands 1–4: 1, poor resolution with a small number of plates (= strong peak distribution); 2, good resolution and a large number of plates; 3, good resolution but a small number of plates; and 4, poor resolution with too small a retention.

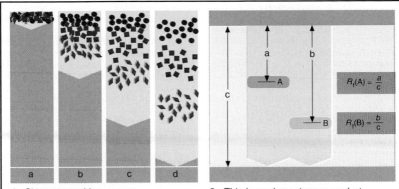

1. Chromatographic process

2. Thin-layer chromatogram product

$$R_f(A) = \frac{a}{c}$$

$$R_f(B) = \frac{b}{c}$$

A. Kinetic theory

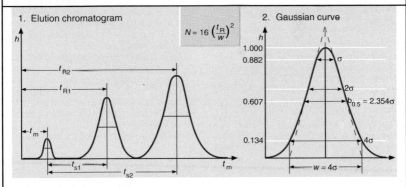

1. Elution chromatogram

$$N = 16 \left(\frac{t_R}{w}\right)^2$$

2. Gaussian curve

$$b_{0.5} = 2.354\sigma$$

$$w = 4\sigma$$

B. Theoretical plate model

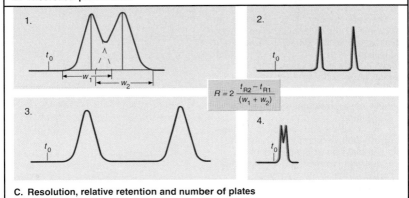

$$R = 2\,\frac{t_{R2} - t_{R1}}{(w_1 + w_2)}$$

C. Resolution, relative retention and number of plates

D. Forms of chromatography. Different forms or shapes of substance peaks are produced following a chromatographic process, depending on the form of the distribution (adsorption) isotherms (see G)—linear or non-linear function—and on the type of chromatography (ideal or non-ideal). With linear, ideal chromatography, which cannot be achieved in actual practice, the retention time is based solely on the distribution ratio and the volume ratio of both phases. The form of the peaks is not changed compared with the form at the start.

1. Linear, non-ideal chromatography. Here the substance peaks broaden out almost symmetrically (Gaussian curve as the ideal shape), as was discussed using the model of the height equivalent to a theoretical plate (B). The distribution isotherm proceeds linearly.

2. Non-linear, ideal chromatography. Here a peak migrates faster the higher local concentration is. Thus, one obtains an asymmetric peak with tailing, since the elution is the weakest where there is a low concentration in the stationary phase (a), therefore at the rear side of the substance zone. If the slope of the distribution isotherm lies above the line, this is described as fronting, or asymmetry before the peak maximum (b).

Non-linear, non-ideal chromatography can be considered to be approximately a linear type for low concentrations (see 1).

E. Dynamic theory. This represents an expansion of the theoretical plate model. Ideal conditions for achieving a low height equivalent to a theoretical plate, but which cannot be realized experimentally in every respect, are as follows:

—an immediate (uninhibited) establishment of equilibrium of adsorption or distribution
—a linear adsorption region or a linear distribution isotherm (see D and G)
—a constant rate in the mobile phase
—a constant temperature in the entire region of the stationary phase
—negligible diffusion.

If no diffusion effects manifest themselves (practically unachievable), this is ideal chromatography.

The broadening of the peaks with the length of the migration distance can be traced back to diffusion, which cannot be overlooked in either the mobile or the stationary phase (v_0 = sample volume at the start of the chromatographic separation procedure).

F. Van Deemter equation. The van Deemter equation describes the association between the height equivalent to a theoretical plate and the dynamic phenomena. A small plate height and therefore a large number of plates for a separation path of a particular length is achieved if the individual terms are as small as possible.

Term A takes into account the scatter diffusion (eddy) and is determined by the type of packing (as homogeneous as possible with particles having a small and uniform size: small term A).

Term B takes into account the diffusion in the longitudinal direction of the separation path (longitudinal diffusion).

Term C includes the rate of equilibrium adjustment, taking into account any interference in the equilibrium between the stationary and mobile phases, C_{stat} and C_{mobile}, respectively, and is called the mass transfer term.

The function of the theoretical plate height as a factor of the linear flow rate is described by a hyperbola with a minimum—the optimum flow rate at the smallest plate height.

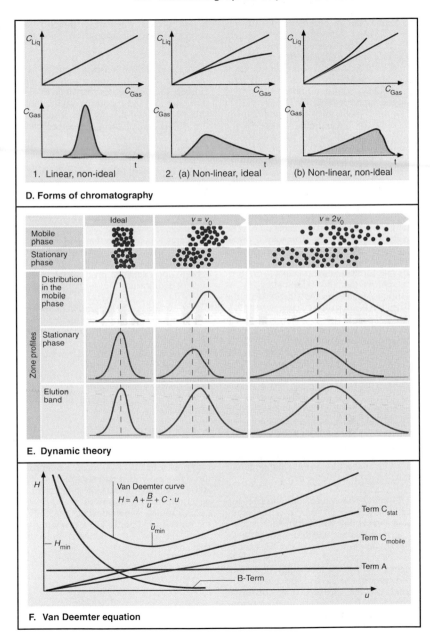

D. Forms of chromatography

E. Dynamic theory

F. Van Deemter equation

G. Separation mechanisms. *1. Adsorption isotherms.* Adsorptions are boundary layer reactions between a gaseous or liquid substance and a solid phase (the adsorbent), which leads to an enrichment of material at the phase boundary layer. The heterogeneous adsorption equilibria are described using adsorption isotherms. The Freundlich adsorption isotherm only satisfactorily describes the linear initial course of the curve ($x/m = ac^{1/n}$, where x is the amount adsorbed, m is the total amount of adsorbent, c is the concentration of the solution and a and n are substance-specific, temperature-dependent constants). The Langmuir adsorption isotherm [$x/m = a \cdot c/(c + b)$] requires a homogeneous surface with monomolecular coating by the adsorbed substance. The adsorption isotherm of Brunauer, Emmet and Teller (c) is valid for multi-layer adsorption. Linear regions are a prerequisite for symmetrical peaks in adsorption chromatography. The strength of retention depends on the steepness of the adsorption isotherms.

2. Distribution isotherms. For distribution, this system consists of two practically immiscible fluids and a third substance soluble in both phases. The separation is based on the different solubilities of the substance in the two liquids. The distribution coefficient α of a substance is the equilibrium constant of a distribution equilibrium. If the latter is independent of the concentrations in both phases, this is a Nernst distribution. Positive deviation from the Nernst distribution in chromatography shows up as fronting (see D) and negative deviation as tailing of the peak.

3. Reversed-phase matrix. In contrast to the usual adsorption system, in a reversed-phase system, the stationary phase is nonpolar (hydrophobic) and the mobile phase is polar (hydrophilic). Silanol (SiOH) groups of silica gel are reacted with a silanization reagent (such as dichlorosilanes) so that hydrocarbon chains can be chemically bound on the surface. Two different mechanisms lead to separations: partitioning between the mobile and the chemically bound phases (the alkyl moiety) and reversed-phase adsorption (on a nonpolar matrix).

4. Ion-pair system. Ion-pair partitioning is based on the formation of an ion pair between the ion to be separated (e.g. from an inorganic ion) and a quaternary ammonium salt, which can be dissolved or adsorbed in the stationary phase (a reversed-phase material). This process is called a dynamic ion-exchange model.

5. Ion exchange. Ion exchange is a stoichiometric partitioning between a solution of ions and a solid substance, to which ions can be bound due to electric forces. Ions with the same charge are reversibly exchanged between the two phases. Cation exchangers contain exchangeable groups such as sulfonic acid groups and anion exchangers have quaternary ammonium groups. Ion-exchange processes lead to a state of equilibrium.

6. Ion exclusion. The separation material in this case is a completely sulfonated cation exchanger where the sulfonic acid groups are hydrated as the mobile phase with water. This hydrate shell is restricted by a partially negatively loaded membrane through which only non-dissociated substances (e.g. carbonic acids with dilute sulfuric acid as the eluate) can pass (inorganic anions are excluded).

7. Gel permeation. Gel permeation is a process by which a distribution according to molecular size occurs based on a sieve effect as a result of molecules migrating through a gel with a porous structure. The steric coupling of certain substances is accomplished by means of the differing accessibilities of individual pore regions in the porous solid body. The chromatogram shows the association between molecular size and elution volume.

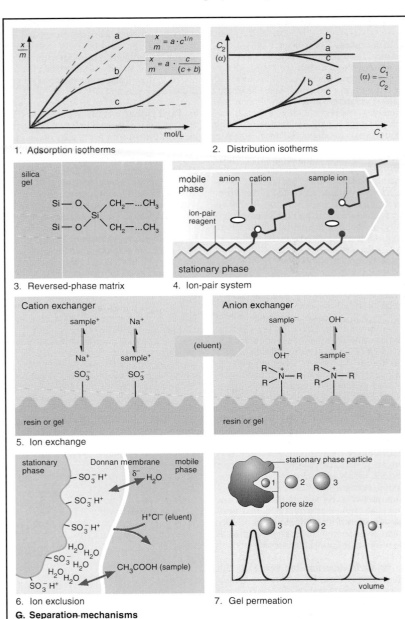

1. Adsorption isotherms

$$\frac{x}{m} = a \cdot c^{1/n}$$

$$\frac{x}{m} = a \cdot \frac{c}{(c + b)}$$

2. Distribution isotherms

$$(\alpha) = \frac{C_1}{C_2}$$

silica gel

Si — O   CH$_2$ — ...CH$_3$
          Si
Si — O   CH$_2$ — ...CH$_3$

3. Reversed-phase matrix

mobile phase   anion   cation   sample ion

ion-pair reagent

stationary phase

4. Ion-pair system

Cation exchanger

sample$^+$ Na$^+$

Na$^+$ sample$^+$

SO$_3^-$ SO$_3^-$

resin or gel

(eluent)

Anion exchanger

sample$^-$ OH$^-$

OH$^-$ sample$^-$

R   R
  N$^+$— R   N$^+$— R
R   R

resin or gel

5. Ion exchange

stationary phase   Donnan membrane   mobile phase

SO$_3^-$ H$^+$ δ^- H$_2$O

SO$_3^-$ H$^+$

SO$_3^-$ H$^+$ H$^+$Cl$^-$ (eluent)

H$_2$O H$_2$O
SO$_3^-$ H$_2$O
H$_2$O H$_2$O   CH$_3$COOH (sample)
SO$_3^-$ H$^+$

6. Ion exclusion

stationary phase particle

1   2   3

pore size

3   2   1

volume

7. Gel permeation

G. Separation mechanisms

Thin-Layer Chromatography

A. Adsorption thin-layer chromatography. The separability of materials based on adsorption processes (p. 146) is determined by the differences in the slopes of the adsorption isotherms. The chromatographic separation process occurs in a thin interlayer. The particle sizes of the separation materials usually lie between 0.5 and 25 μm. They are deposited from aqueous suspensions in thin layers 0.5 to 0.25 μm thick on glass, aluminum or plastic plates by spraying, pouring or, preferably, using spreaders. Today they are usually commercially available as coated plates (prepared plates). If the adsorption isotherms are flat, substance B will be found principally in the mobile phase; if the slope increases it is found more in the stationary phase (substance A migrates only a short distance from the starting point).

B. Elution activity of different solvents. A suitable solvent/mobile phase is selected based on the elution strength: the solvents are listed in order of increasing eluting effect in series of elutions. A rapid procedure can be used to make an initial selection of a mobile phase for the separation of a mixture of substances. The mixture is applied multiple times on a TLC plate, each addition with a drop of the solvent. Benzene is best suited for separation with the sample shown.

C. Elution strength in a solvent mixture. The elution strength (or power) is a measure of the elution capability of a solvent; it is correlated with the polarity of the eluent. Besides simply changing the solvent, the elution strength can also be changed by using mixtures. By adding the polar component diethyl ether to the nonpolar pentane, the elution strength increases rapidly at first, then the slope decreases to approach the value for the pure polar solvent.

D. Triangle diagram for selecting the adsorption milieu. By adsorption milieu we mean the combination of absorbent and eluent ('strict' milieu: a combination of aluminum oxide and hexane with pre-dominant adsorption; see A). A triangle diagram is used to select the eluent and the activity of the adsorbent for adsorption chromatography for a mixture with known polarity. The activity of adsorbents is divided into five categories (using dye test mixtures).

E. Performing a TLC separation. The mixture to be separated is applied approximately 1 cm from the lower edge of the TLC plate, e.g. in the form of a dot using a glass capillary. Then the TLC plate is placed in a developing chamber filled with less than 1 cm of the solvent and with paper attached to the back wall (to saturate the atmosphere with the vapor of the solvent). The mobile phase starts to migrate upwards.

F. Development chambers. Only a single TLC plate can be placed in a flat-bottomed chamber (1), but a double-trough chamber (2) can hold two TLC plates at a time. The R_f values are decisively influenced by the degree of saturation of the chamber with the solvent. If the chamber atmosphere is not saturated, the solvent will evaporate from the plate during the chromatographic process. A sandwich chamber (3) has only a very small chamber volume, owing to the spacer: if it is moistened, the chamber is saturated. Flat-bed chambers (4) are used for high-performance TLC (HPTLC). The plates, lying with the TLC layer facing down, are developed from two sides. The solvent is transported to the layer using a glass plate. When the solvent fronts meet in the middle, a straight line is formed. Solvent amounts and chamber volumes are small and easily controlled.

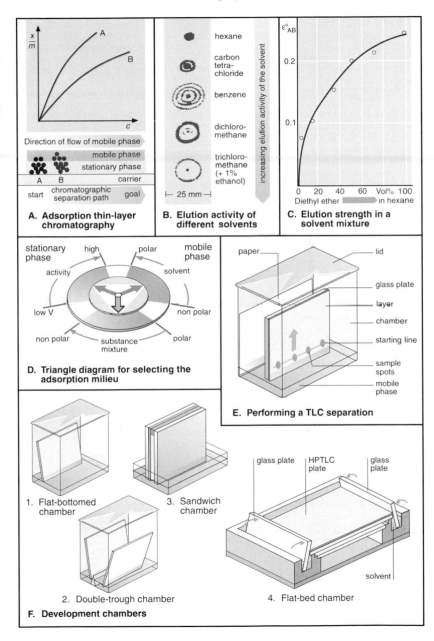

A. Adsorption thin-layer chromatography

B. Elution activity of different solvents

C. Elution strength in a solvent mixture

D. Triangle diagram for selecting the adsorption milieu

E. Performing a TLC separation

F. Development chambers

1. Flat-bottomed chamber
2. Double-trough chamber
3. Sandwich chamber
4. Flat-bed chamber

G. R_f, R_{st} and R_m values. The qualitative evaluation of chromatograms in TLC (inner chromatograms, since the substances remain within the separation path) is done by determining the delay of the substance transport compared with the solvent front. The retention factor, R_f, is the ratio of the distance of the substance (A) from the starting line (a) and the distance of the solvent front from the start (c): $R_f = a/c$. Correspondingly, the R_f value for a standard (reference) material as R_{st} is calculated as a/s. In referring to the R_{st} values, the R_f value or a standard substance is arbitrarily set equal to 1. For homologous series such as aliphatic carbonic acids, an R_m value is defined as the logarithm of the product of a constant and the distribution coefficient for a substance between two phases. R_m values generate a straight line for a homologous series as a function of the number of carbon atoms.

H. Development techniques. The typical method for development is ascending TLC (1 and 2) with a dotted or line-shaped application of the mixtures. In the step method, to improve the separations one allows several solvents with differing elution strengths to migrate one after another in the same direction across the layer. If one applies the same mobile phase multiple times one after another, this is called multiple TLC. The tapered strip technique (3) also provides improved separations compared with the conventional method, as does the circular method (4). In both of these techniques, the solvent runs not only in the direction of movement, but also perpendicular to it. The substances are pulled from one another lateral to the direction of migration, and they do not overlap even if there are only slight differences in their R_f values. In the circular method, the solvent flows through a drilled hole over a wick in the middle of the TLC plate, whereby the mixture can be applied in a concentric ring. With anticircular development (5), on the other hand, the substances are located on an outer circular line; the mobile phase is applied from the outside in a circular fashion, and it flows inward. This reduces the enlargement of the spots due to transverse diffusion with substances having high R_f values. In two-dimensional TLC (6), the mixture for the first run is applied in a corner of the plate (6.a) and, after development in the first direction, it is dried. Then one rotates the plate by 90° so that the partially separated mixture is located at the lower edge of the plate, and one develops the plate using a second solvent in the second direction (6.b).

I. Spot penetration depth as a factor of the R_f value. When a mixture is applied to the starting point of a TLC plate, the substances penetrate into the layer. When the plate is 'developed,' the evaporation of the solvent from the surface of the plate (especially in non-saturated chambers; see E and F) creates a flux by the mobile phase, which pushes the spots to the surface. This fact is particularly important for quantitative evaluations.

J. Automatic sample application. A microliter dosing syringe can be used to apply very uniformly volumes between 5 and 20 (up to a maximum of 100) μL linearly using the appropriate device. These devices are controlled by a microprocessor and are programmable.

K. Immersion chamber for applying reagents. Substances can be made visible without a chemical reaction due to their own coloring, UV absorption or fluorescence. In addition, substances or groups of substances can be made visible using characteristic coloration after reaction with sprayed reagents. Instead of spraying, often it is recommended that one use an immersion chamber, whereby the reagent is distributed more uniformly (important for quantitative analysis).

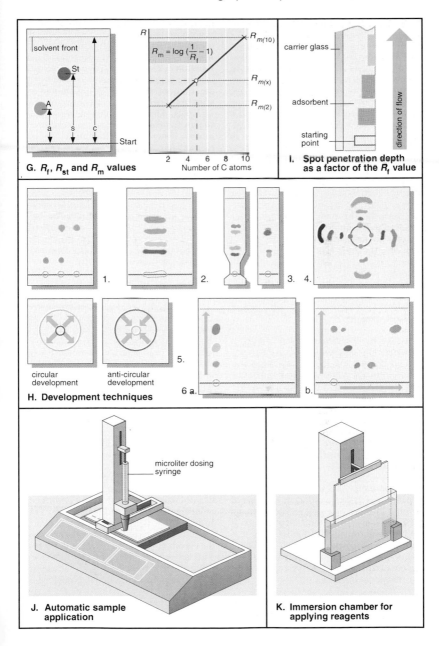

G. R_f, R_{st} and R_m values

$R_m = \log \left(\frac{1}{R_f} - 1 \right)$

Number of C atoms

I. Spot penetration depth as a factor of the R_f value

carrier glass

adsorbent

starting point

direction of flow

H. Development techniques

circular development

anti-circular development

J. Automatic sample application

microliter dosing syringe

K. Immersion chamber for applying reagents

L. Quantitative thin-layer chromatography.

1. Measuring principles of a TLC scanner.

TLC scanners for the quantitative evaluation of thin-layer chromatograms consist of a spectral photometer, a mechanical (programmable) device for transporting the TLC plate through the light beam coming from the photometer's light source and a recorder for recording the remission–position curves. An absorption measurement in incident light is called a remission measurement (remission = diffuse reflection). The monochromatic light emanating from the light source is beamed on to the TLC plate at a particular angle, and the reflected light is measured in remission in the receiver. Since the TLC plate continues to move in the flow direction of the mobile phase during the measurement, one records a remission–position curve. With a substance-free layer, the light is largely reflected without significant absorption. Depending on the structure of the compound (chromophore) and its concentration in the layer, substance spots absorb part of the light energy.

2. Comparison of visual and densitometric thin-layer chromatograms.

Semi-quantitative evaluation of thin-layer chromatograms is possible without any equipment by measuring the area of the separated substances (a). Densitometers work similarly to photometers: light passing through the TLC layer (on glass plates), or its intensity, is compared with that of the incident light. In this manner a chromatogram is produced as a TLC densitogram (b), as with remission measurements.

3. Quantitative TLC analysis of quinine.

Separations for the quantitative analysis of quinine in a tincture of cinchona can be performed in 30 min on silica gel (solvent = dichloromethane–diethylamine) with a length of run of 15 cm. Standard solutions with a known quinine level (1–3) can be used in the TLC analysis (at 320 nm) in a tincture of cinchona.

M. Calibration functions.

The degree of remission, the portion of the radiated light intensity which is reflected, or not absorbed, is not correlated in a linear fashion with the concentration. In many cases, there is a linear correlation using the Kubelka–Munk function: $(1 - R_{abs})^2/2R_{abs} = Kc$ (R_{abs} = absolute degree of remission; K = constant, the ratio of the spectral absorption and scattering coefficients; c = concentration in μg/spot). The three calibration functions shown for the dye Sudan red were determined using the following evaluation methods: in curve (1) there were point-by-point remission measurements at 500 nm and an evaluation using the Kubelka–Munk function (as a surface integral). Curve (2) is also linear, and in this case the remission is also measured, but it is evaluated according to the surface area squared. Transmission measurements with a determination according to the area (in absorption units) does not provide a linear calibration function (3).

N. Principle of fluorescence reduction.

Substances which are colorless but which absorb in the UV region can be detected on TLC layers if a fluorescence indicator (manganese-activated zinc silicate, pyrene derivatives, morin, etc.) is added to the substance. In general, fluorescence is excited at 250 or 280 nm. If there are substances on the layer which can absorb light in this region (especially aromatic compounds), the fluorescence is reduced. The substances appear as dark spots on a green fluorescent layer.

O. Comparison of TLC and HPLC.

Thin-layer chromatography has developed into a instrumental method today which still plays an important role even though HPLC (p. 156) is available. In particular, separations of substances with strongly differing polarities are performed better using TLC. In addition, with LC several samples can be separated at the same time. Comparison of the chromatograms shows the efficiency of TLC (N = adsorption TLC, R = reversed-phase TLC) compared with adsorption HPLC for polychlorinated naphthalenes.

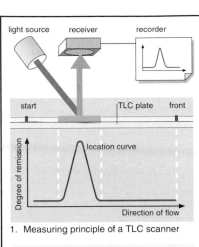

1. Measuring principle of a TLC scanner

a. Visual thin-layer chromatogram

b. Densitogram
separation time
35 min

2. Comparison of visual and densitometric
thin-layer chromatograms

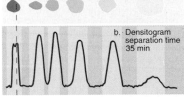

1	375 ng quinine
2	750 ng quinine
3	1500 ng quinine
4	tincture of cinchona with 750 ng quinine (= 0.075% quinine)

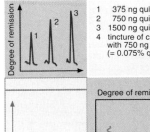

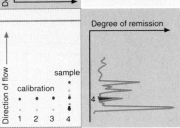

3. Quantitative TLC analysis of quinine

L. Quantitative thin-layer chromatography

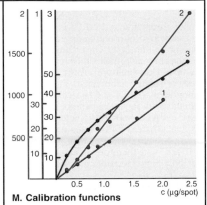

M. Calibration functions

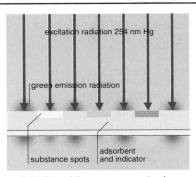

N. Principle of fluorescence reduction

excitation radiation 254 nm Hg

green emission radiation

substance spots   adsorbent and indicator

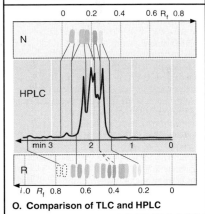

O. Comparison of TLC and HPLC

Column Liquid Chromatography

Liquid chromatography (LC) in columns, also known as liquid chromatography or–in contrast to thin-layer chromatography–as column chromatography, differentiates the classical (low-pressure) from the high-pressure (high-performance) liquid chromatography (HPLC). Classical LC uses columns with inner diameters of 1 cm (or more) and particles with diameters of 100 to 200 μm, whereas HPLC uses narrow columns (2–4 mm diameter) with particles of 3–10 μm, and therefore it achieves considerably higher separation efficiency.

A. Principles of partition chromatography. *1. Matchbox model.* The transition of an extraction process into a partition chromatographic separation can be modeled as a series of successive partitioning steps. Each step corresponds to adjustment of the equilibrium between the stationary and mobile phases. In this example, substances A and B partition themselves by 50 and 75% in phase 2 (the mobile phase). After the stepwise continuation of phase 2, a new partition equilibrium always established itself between the two phases. An apparatus which works based on this model is called the Craig countercurrent apparatus and it is used for the separation of natural substances.

2. Partitioning after five separation steps. After the conclusion of five separation steps, the portions for substances A and B are added up in the boxes located above one another, and they are displayed in the form of a distribution curve. In the ideal case, the distribution across the separation path (as a function of the box number in this case) corresponds to two Gaussian curves. The model makes it clear that as the number of separation steps increases with a given distribution, one can achieve an added improvement in the separation.

3. Stepwise development of a separation. If the theoretical plate number is increased by lengthening the column, i.e. the separation path, then on the one hand there is a broadening of the substance bands (as a result of

diffusion effects), and on the other hand there is also an improvement in the separation of the two substances A and B. The figure shows the development of a separation along the length of a separating column.

B. Liquid (column) chromatography. The glass column with a stopcock is filled with the selected separation material. Fine particles are prevented from flowing through by a wad of glass-wool before the outlet. The sample is applied at the uppermost part of the column and the eluent flows through the column, in this case at atmospheric pressure. The eluate, which contains the substances dissolved in the mobile phase, is collected (often in fractions).

1. Displacement technique. The displacement technique can be successfully applied if substances are retained very strongly in the stationary phase, so that large volumes of mobile phase would be needed for the transport through the column. To develop a displacement chromatogram, a substance A is added to the mobile phase; this substance is retained more strongly by the stationary phase than is substance B. Substance A displaces B from the stationary phase and pushes it in front of itself. Usually only incomplete separations can be attained.

2. Elution technique. Elution is the process where the mobile phase flows until all substances to be separated have left the column.

3. Elution procedures in the column. Elution is based on the interactions between the substances to be separated and the stationary and mobile phases. It can also be understood to be the elution of the materials from the stationary phase. After the sorption step, in each separation step there is an adjustment of the equilibrium between sorption and elution. The progressive elution leads to an increased separation of the substances.

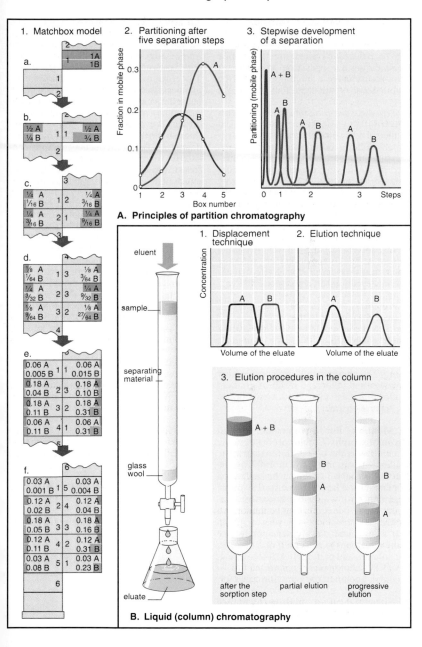

1. Matchbox model

2. Partitioning after five separation steps

3. Stepwise development of a separation

A. Principles of partition chromatography

B. Liquid (column) chromatography

eluent

sample

separating material

glass wool

eluate

1. Displacement technique

2. Elution technique

3. Elution procedures in the column

after the sorption step

partial elution

progressive elution

D. Gradient elution. If different substances have very different retention times, the analysis time for the entire separation can be reduced by changing the mobile phase during the separation procedure. A continuous change in the composition of the mobile phase is called gradient elution.

1. Low-pressure gradient system. Solvents which differ in polarity are transported out of the solvent vessels L_1 to L_3 using regulatable pumps, through the valves or low-pressure pumps V_1 to V_3 to produce a mixture in the mixing chamber which continuously changes its make-up, and from there the mixture is transported to the separation column using another pump.

2. Binary gradients. As a rule, concentration gradients are formed with a stronger eluting solvent (B) in more weakly eluting solvent A. The composition (%B in A) as a function of time can produce a curve that is linear, stepped linear, convex or concave.

3. Comparison of isocratic and gradient elution. Although the three substances 1, 2 and 3 can be separated isocratically using a single eluent which does not change its composition, substance 3 does have a long retention time and marked band broadening. The elution bands are considerably narrower and taller as a result of gradient elution; this is associated with a reduction in the separation time.

E. Modules for low-pressure liquid chromatography. Liquid chromatography in columns is the oldest chromatographic method; it was first used in 1903 by the Russian botanist Tswett for the separation of plant dyes. The components of a simple apparatus for low-pressure liquid chromatography include: the eluent container as a storage vessel for the mobile phase; a pump (e.g. a hose pump); a valve for applying the sample, perhaps with the aid of a syringe; the column; and preferably a detector (e.g. UV/VIS photometer) connected to a recorder, a computer or a fraction collector.

F. Structure of a high-performance (high-pressure) liquid chromatograph.

An HPLC device consists of four main components: a pump, an injection system, the separation column and the detector with a processing system. HPLC uses separation particles with a particle size of 3 to 10 μm. Therefore, it achieves high theoretical plate numbers (see the van Deemter equation), but at the same time it makes it necessary to overcome a relatively high counterpressure during the transport the mobile phase through the narrow column (2–6 mm inner diameter). All components must be connected to one another so as to be completely free of dead volume if possible (capillary attachments are used), and they must be resistant to pressure (up to approximately 300 bar or 30 MPa). When the sample is applied, it is first injected without pressure into a sample loop located in a four-way valve. By switching, the eluent flow goes through this sample loop, whereby the sample is brought into the column. The analytical separation column, usually made of stainless steel, should be thermostatable. For detection, one uses UV/VIS and fluorescence spectrometers, refractive index and amperometric and conductivity detectors with small flow-through cells (a few microliters in volume).

G. Reciprocating pump. *1. Functional diagram.* High-pressure pumps must ensure a constant and pulse-free flow of the mobile phase. With reciprocating pumps consisting of an eccentric attachment with variable speed and a piston made of sapphire driven by the pump, the aspirated liquid is expelled via the forward movement of the pump head and pushed into the column via ball valves made of sapphire.

2. Flow profile. As a result of this process, undesirable pulsations occur with a single-piston pump (2.a); these can be eliminated by using a double-piston pump (2.b) with active damping.

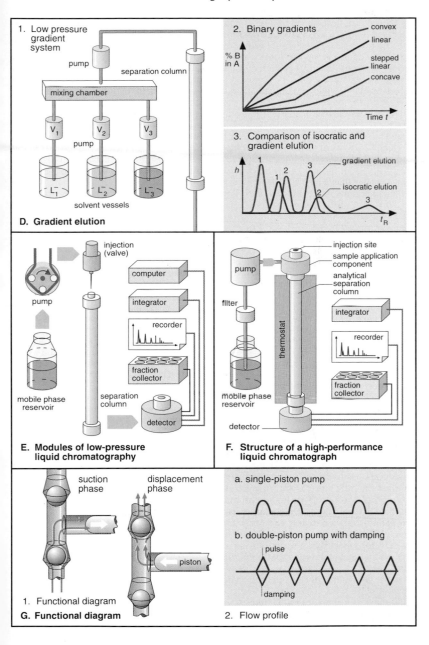

1. Low pressure gradient system

pump

separation column

mixing chamber

V_1 V_2 V_3

pump

L_1 L_2 L_3

solvent vessels

D. Gradient elution

2. Binary gradients

convex
linear
stepped linear
concave

% B in A

Time t

3. Comparison of isocratic and gradient elution

h

1
1   2
3

gradient elution

isocratic elution

2

3

t_R

injection (valve)

computer

integrator

recorder

fraction collector

pump

mobile phase reservoir

separation column

detector

E. Modules of low-pressure liquid chromatography

injection site

sample application component

analytical separation column

pump

filter

thermostat

integrator

recorder

fraction collector

mobile phase reservoir

detector

F. Structure of a high-performance liquid chromatograph

suction phase

displacement phase

piston

1. Functional diagram

G. Functional diagram

a. single-piston pump

b. double-piston pump with damping

pulse

damping

2. Flow profile

H. Sample application valve. The sample is applied using six-way valves. In position a the mobile phase flows via 2/3 directly into the column. The sample loop, a capillary with volumes usually between 100 and 200 μL, can be filled with a microliter syringe. By switching the valve into position b (2/1/4/3), the sample is transported into the HPLC column with the help of the eluent.

I. Height equivalent to a theoretical plate and particle diameter. The high efficiency of HPLC lies in the use of particles with diameters less than 10 μm. The dependence of the height equivalent to a theoretical plate on the square of the average particle diameter is derived from the differentiated van Deemter equation (see p. 144). For diameters smaller than 80 μm, experimentally one finds a smaller proportionality between $d_p^{1.3}$ and $d_p^{1.8}$; for $d_p < 3$ μm, H is even proportional to d_p. In actual practice, separating materials of 3–10 μm diameter are used; they are used to fill stainless-steel columns with threads and grids or sieves on the ends which have a low dead volume.

J. Flow-through cell of a detector. The end of the column is connected to the detector also using capillaries and threads having little or no dead volume. To avoid turbulence and mixtures of separate substances, a Z-shaped flow-through cell can be used for absorption measurements of the light at a particular wavelength.

K. Detector noise and drift. The basic requirements for a detector can be summarized in three points: it should have a high measuring sensitivity, little variation in the baseline and little noise. Drift is the deviation of the zero line from a straight line parallel to the time axis. Changes in the zero line from a straight line over a certain time are called noise. Sources of noise include the detector's electronics, variations in the surrounding temperature and gas bubbles in the eluent flow or changes in the flow-through (pulsation of the pump). The detector noise level determines the detection limits of an analytical procedure. As a rule, only signals which exceed a specific

(3–6-fold) multiple of the noise are identified as such.

L. Refractive index (RI) detector. The RI detector is a universally applicable, but relatively insensitive detector which records changes in the refractive index. If a substance is in the eluent flow, a change occurs in the refractive index. RI detectors require constant temperatures and flow rates. They are mostly used in situations where the separated materials do not exhibit any UV absorption (such as with sugars). A differential refractometer consists of a prism and a variable base plate with a measuring cell. The irradiating light dissociates into a prism-reflecting portion and another portion which proceeds into the measuring cell and from there to the photodiode (after reflection).

M. Photodiode array detector. In contrast to the typical UV detectors (for measuring the absorption of light in the UV region), a photodiode array detector containing a large number of small photodiodes can be used to make concurrent measurements at different wavelengths. In this way one obtains a complete spectrum at each point in time of a chromatogram, and with a computer evaluation system one can generate 3-D chromatograms.

N. Amperometric detectors. Glassy carbon is used as an electrode material (working electrode) in flow-through cells in connection with an auxiliary electrode and a reference electrode for the anodic determination (oxidation) of organic substances. Depending on the type of construction, one differentiates between the wall jet cell 1 and the thin-layer cell 2. The current intensity is measured as a function of time with a constant potential of the working electrode.

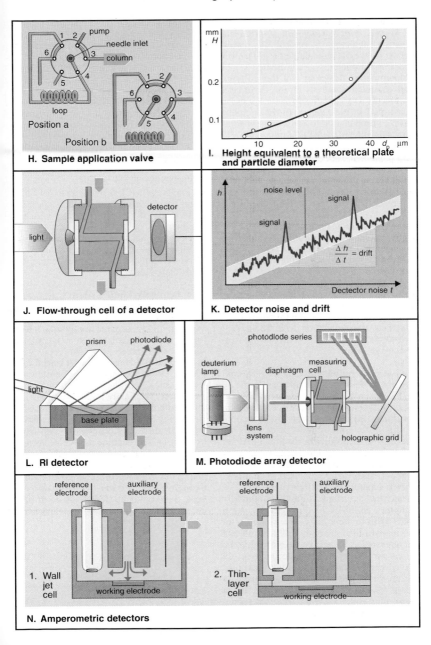

H. Sample application valve

Position a
Position b

I. Height equivalent to a theoretical plate and particle diameter

J. Flow-through cell of a detector

K. Detector noise and drift

$$\frac{\Delta h}{\Delta t} = \text{drift}$$

L. RI detector

M. Photodiode array detector

N. Amperometric detectors

1. Wall jet cell
2. Thin-layer cell

O. Chemical reaction detector.

1. System configuration. In a chemical reaction detector, a reactor is placed between the actual detector (UV/VIS spectrophotometer, fluorimeter or amperometric detector) and the column, and chemical reactions of the chromatographically separated substances can be performed in the reactor. In the simplest case, a pump is used to add a reagent solution at the inlet of the HPLC column via a T-piece as a mixing chamber. The reactor can be a plastic tube several meters long which is wrapped about a glass pipe. The mixing chamber and the reactor must be designed so that no significant diffusion effects, and therefore little band broadening, can occur.

2. Determination of thiamine. The selectivity in a detection process can be enhanced considerably using this type of post-column derivatization, as shown in the determination of vitamin B_1 (thiamine) after oxidation with hexacyanoferrate(III) and fluorescence detection in a food sample. Chromatogram (a) shows the application of a UV detection at 254 nm; chromatogram (b) was obtained after oxidation to a fluorescent derivative, thiochrome, using a fluorescence detector.

P. Ion chromatography. In ion chromatography, described for the first time in 1975, an ion-exchange separation column was coupled with a suppressor column and a conductivity detector for on-line detection. In this system, the suppressor column has the task of neutralizing the eluent, or of largely eliminating the background conductivity of the mobile phase. A mixture of Na_2CO_3 and $NaHCO_3$ is used as an eluent for the separation of anions (with very low concentrations of a few mmol/L). In early studies, a cation exchanger in the H^+ form was used to neutralize the eluent after separation. Today, hollow-fiber membrane suppressors with sulfonated polyethylene fibers are used for the same purpose; an exchange of sodium and hydrogen ions (from dilute sulfuric acid) takes place in these fibers following the countercurrent principle. This makes the frequent regeneration of the ion exchanger unnecessary.

Thus only hydronium ions and sodium anions are transported into the conductivity detector with anion separation; these induce a large change in the conductivity.

Today, the term ion chromatography includes all rapid liquid chromatographic systems for the separation of inorganic and organic substances which can dissociate into cation or anions.

Ion analyses have been performed with conductometric detection even without a suppressor system since 1979, where ion exchangers having limited exchange capacity are used with eluents having limited conductivity. There are different types of ion exchangers (2): (a) those made of a film of an ion exchanger on an inert core; (b) those made of a porous polymer which is polymerized on an inert surface such as silica gel; (c) those made of a macroporous polymer; and (d) those consisting of an anion exchanger bound electrostatically to a sulfonated particle.

Q. Chemically bonded phases. Silica gels with surface OH groups (as silanol —Si— OH groups) can be made to react with a monofunctional organic compound, which brings about monolayer coverage of the surface with groups of organic molecules. In this way alkyl groups can be linked with cyano-, phenyl- or amino-terminal groups on the silica gel. These separation phases are called chemically bonded phases. The different polarities of such phases are made clear using the example of the separation of different aromatic compounds with tetrahydrofuran and heptane as eluents, as normal-phase chromatography on chemically modified silica gels. The advantage of these phases lies in the fact that changes in activity through traces of water in the organic solvents do not play a role here, in contrast to the adsorption processes on unmodified silica gel.

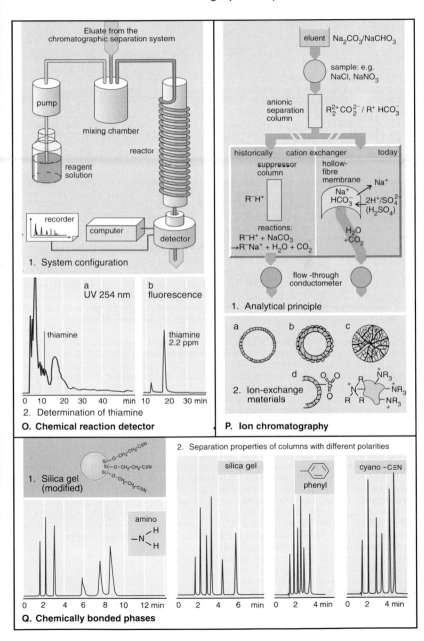

Eluate from the
chromatographic separation system

pump

mixing chamber

reactor

reagent
solution

recorder

computer

detector

1. System configuration

a
UV 254 nm

thiamine

b
fluorescence

thiamine
2.2 ppm

0 10 20 30 40 min   10   20   30 min

2. Determination of thiamine

O. Chemical reaction detector

eluent $Na_2CO_3/NaCHO_3$

sample: e.g.
NaCl, $NaNO_3$

anionic
separation
column $R_2^{2+} CO_2^{2-} / R^+ HCO_3^-$

historically   cation exchanger   today

suppressor
column

hollow-
fibre
membrane

R^-H^+

Na^+
HCO_3^-

Na^+

$2H^+/SO_4^{2-}$
(H_2SO_4)

reactions:
$R^-H^+ + NaCO_3$
$\rightarrow R^-Na^+ + H_2O + CO_2$

H_2O
$+CO_2$

flow-through
conductometer

1. Analytical principle

a   b   c

d

2. Ion-exchange
materials

N
R
R

NR_3^+

NR_3^+

NR_3^+

P. Ion chromatography

Si—O-CH$_2$-CH$_2$-C≡N
Si—O-CH$_2$-CH$_2$-C≡N
Si—O-CH$_2$-CH$_2$-C≡N

1. Silica gel
(modified)

amino

$-N\begin{smallmatrix}H\\H\end{smallmatrix}$

0 2 4 6 8 10 12 min

2. Separation properties of columns with different polarities

silica gel

phenyl

cyano –C≡N

0 2 4 6 min

0 2 4 min

0 2 4 min

Q. Chemically bonded phases

R. Reversed-phase chromatography. In reversed-phase chromatography, the stationary phase is less polar than the mobile phase.

1. Surface of a reversed phase. Surface-modified silica gels can be obtained by reacting silanol groups with alcohols (see Q). A greater hydrolytic stability of the binding is achieved via the reaction with alkylchlorosilanes, whereby Si—O—Si compounds are formed. The groups chemically bound to the silica gel surface are also called hydrocarbon brushes. In reversed-phase systems, substances are retained by the non-polar stationary phase all the more, the less water-soluble or the less polar they are. An elution series with decreasing retention can be set up as follows: aliphatic compounds—induced and permanent dipoles (chlorinated hydrocarbons)—Lewis bases (such as ethers, aldehydes, ketones, amines)—Lewis acids (alcohols, phenols, carboxylic acids). In a homologous series, the retention increases with increase in the number of carbon atoms.

2. Comparison of the methyl and octadecyl phases. In a group of substances such as uracil (base), phenol, acetophenone, nitrobenzene and toluene, the retention in this series increases in accordance with the general rule. The selectivity of the separation is also increased considerably with the same composition of the mobile phase of methanol and water via elongation of the carbon chain at the stationary phase.

S. Gel permeation chromatography.
1. Cylinder pore model. Gel permeation is a process whereby a distribution according to molecular size takes place when molecules migrate through a gel, due to a sieving effect. The stationary phase consists of a porous material. Small molecules can penetrate into all pores, so they have the entire volume of the mobile phase in the column available. Therefore, they are the last to be eluted. However, if the molecules are larger, only a portion of the total pore volume is accessible. A coiled molecule with an average statistical radius r_1 can be

retained in a larger pore volume than a molecule with radius r_2.

2. Gel chromatographic separation of styrene oligomers. The first point in a gel chromatogram represents the material that has no access to the pores because of its size. Its elution volume is described as the interstitial volume V_I and is called the exclusion volume (after the signal for $n = 14$ here). The smallest molecule with the signal to the right corresponds approximately to the dead volume portion of the retention volume. These molecules have only migrated by diffusion; compared with the 'excluded' molecules in the separation system, they are retained. The difference between V_I and V_0 corresponds to the pore volume of the stationary phase.

T. Bioaffinity chromatography. In general, affinity chromatography refers to the separation of substances using selective or specific interactive forces. Biochemically, specific interactions between enzymes and substrates or between antibodies and antigens play an important role.

1. Principle. Particular ligands are covalently bound to water-insoluble materials such as the polysaccharide agarose or glass beads, and the ligands in turn can only bind to special substances, spatially and electrostatically. The adsorption from solution is reversible, so it can be undone using a selective eluent.

2. Bioaffinity chromatography of peroxidase. The enzyme peroxidase can be selectively separated from a protein mixture using this method. The elution of peroxidase, a glycoprotein, was started where the arrow is pointing with the aid of a special sugar as a selective eluent. A special lectin (a plant protein) bound to silica gel was used as the selective binding partner.

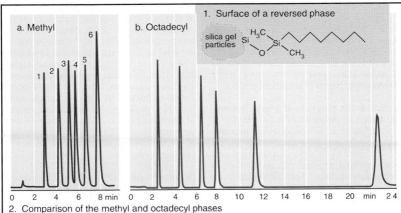

a. Methyl

b. Octadecyl

1. Surface of a reversed phase

silica gel particles

2. Comparison of the methyl and octadecyl phases

R. Reversed-phase chromatography

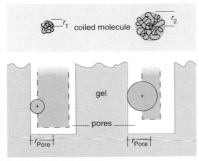

coiled molecule

gel

pores

1. Cylinder pore model

S. Gel permeation chromatography

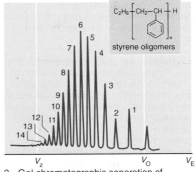

styrene oligomers

2. Gel chromatographic separation of styrene oligomers

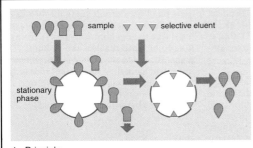

sample   selective eluent

stationary phase

1. Principle

T. Bioaffinity chromatography

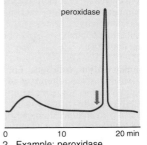

peroxidase

2. Example: peroxidase

Gas Chromatography

In gas chromatography (GC), an inert gas represents the mobile phase as it flows through a tube in which the stationary phase is located. The analytes, in either a gaseous or vapor state, can be separated using adsorption or partition chromatography.

A. Schematic diagram of a gas chromatograph. The structure of a GC device is derived from the sub-steps which proceed continuously one after another. A gas, a volatile liquid or an undecomposed volatile solid is applied to the column via the sample application section. The substances are transported through the thermostatted column with the aid of a carrier gas, and the chromatographic process takes place at a pre-defined temperature there. After leaving the column, one after another the separated substances in the gas phase pass a detector which displays each individual component using a recorder or an integrator.

B. Symbols for GC equipment. The characteristic components of a GC device are taken from the GC analytical procedure—displayed in symbols: pressure and flow regulators, chokes, needle valves, manometers and flow meters serve to provide and control the gas supply, both for the separation column and the sample application component, and for the detector. Different injectors (see E) are used for applying the sample. Packed or capillary columns (see F) are used as separation columns, sometimes along with a filter column. Connected to the thermostatted furnace chamber are detectors such as the HCD (heat conductivity detector), the FID (flame ionization detector) or ECD (electron capture detector).

C. Separation columns. The stationary phase of a packed column (an adsorbent or a liquid phase on an inert carrier material) is located in tubes usually made of glass or quartz with an inner diameter of 3 to 8 mm and lengths of 1 to 3 m. Capillary columns have inner diameters of 0.2 to 1 mm and lengths of up to 100 m or more. With thin-film capillaries, the liquid phase (stationary phase) is directly on the inner wall of the capillary tube ($1-3 \mu$m film); with thin-layer capillaries, the thin layer of a carrier material on the inner wall is coated with the liquid phase (they can withstand higher loads than thin-film capillaries). All three capillary columns have an open longitudinal channel in the capillaries, since there is no packing, whereby the gas volume in the mobile phase is large compared with the liquid phase. Theoretical plate numbers of 100 000 and higher can be attained with capillary columns.

D. Inactivation of surfaces. To prevent undesirable interactions of substances to be separated with the hydroxy groups of the glass surface, a reaction with di- or trimethylsilyl groups is used for inactivation: a hydrophobic surface is formed.

E. Sample injector. Liquid and gaseous samples are introduced through a septum (usually made of silicone rubber) at the head of the column into the carrier gas flow using a hypodermic syringe. With liquids, at the start of the column there is an injection block which can be heated separately from the column furnace in order to achieve an instantaneous transition to the gas phase at higher temperatures than the column temperature (1). Owing to a weaker carrier gas flow (0.2 to 2 mL/min instead of 30–40 mL/min), the injection volume must be considerably lower (0.1 to 1 μL) with capillary columns than with packed columns. A reduction of the sample volume is possible with low-load columns using subdivision of the current (2) (split injection). The splitting ratio can be adjusted using resistance to fluid flow; only part of the total sample volume reaches the column. The higher the choke resistance is set relative to the resistance to fluid flow of the column, the more sample reaches the column. An undivided injection (by closing the split outlet) can also be used in trace analyses (3).

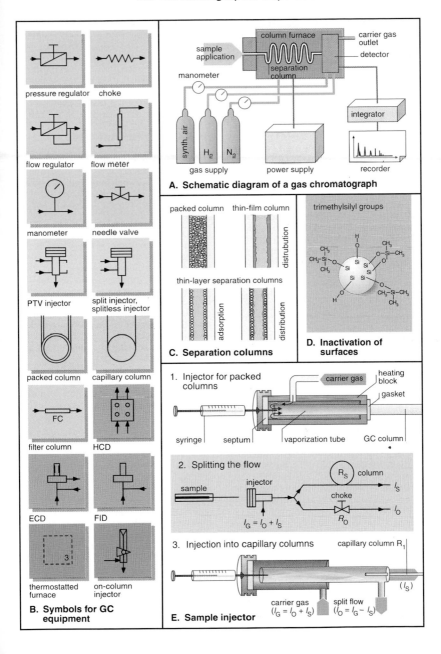

B. Symbols for GC equipment

pressure regulator
choke
flow regulator
flow meter
manometer
needle valve
PTV injector
split injector, splitless injector
packed column
capillary column
FC
filter column
HCD
ECD
FID
thermostatted furnace
on-column injector

A. Schematic diagram of a gas chromatograph

column furnace
carrier gas outlet
sample application
detector
manometer
separation column
synth. air
H_2
N_2
gas supply
power supply
integrator
recorder

C. Separation columns

packed column
thin-film column
distribution
thin-layer separation columns
adsorption
distribution

D. Inactivation of surfaces

trimethylsilyl groups

E. Sample injector

1. Injector for packed columns

carrier gas
heating block
gasket
syringe
septum
vaporization tube
GC column

2. Splitting the flow

sample
injector
R_S column
I_S
choke
R_O
I_O
$I_G = I_O + I_S$

3. Injection into capillary columns

capillary column R_1
(I_S)
carrier gas
$(I_G = I_O + I_S)$
split flow
$(I_O = I_G - I_S)$

F. Comparison of GC separations with packed and capillary columns. The theoretical plate numbers of a capillary column are generally higher than those of a packed column by a factor of 100. As a rule, the height equivalent to a theoretical plate is only smaller by a factor of 2, so the high theoretical plate number comes about due to the length of the column. Among other things, the reduction of the height equivalent to a theoretical plate can be traced back to the absence of a packing term A in the van Deemter equation (see p. 144). Thin-layer columns (SCOT: support-coated open-tubular column) are used almost only for adsorption GC separations. With thin-film columns (WCOT: wall-coated open-tubular column), the liquid phase is applied as a thin film on the inner (roughened) surface of the capillary. WCOT capillary columns with immobilized liquid phase represent an important step forward—owing to a polymeric cross-linkage or by alchemical bonding to the column wall (via the OH groups on the silicon), whereby high-temperature stability (slight bleeding by the column) is attained. The high efficiency of a capillary column is made clear using the example of the separation of a mixture of ethereal oils: substance groups or pairs such as Nos 2/3, 5–7, 8/9, 14/15 from the packed column (1) are separated well in the capillary column (2).

G. Kováts indices. To characterize retention properties, the retention index (Kováts, 1958) was developed in place of the retention time for the homologous series of n-alkanes as standard substances. It indicates the relative position of a substance peak in a chromatogram, whereby one interpolates between the retention times of the neighboring n-alkanes. For the logarithm of the retention time, $\log t_s$, of the homologous series, at first there is a linear dependence on the number of carbons—to be understood from the thermodynamic interpretation of the vaporization process. The retention index is dependent on the type of the substance to be partitioned, the temperature during the chromatographic separation process (slight and linear) and the type of stationary phase. It is defined by

$I = 100(y - x)$ $(\log t_{sA} - \log t_{sx})/(\log t_{sy} - \log t_{sx}) + 100x$, where x and y are the carbon numbers of n-alkanes before and after substance A with the corresponding net retention times. Curve 1 shows the linear correlation between the logarithm of the net retention time t_s and the number of carbon atoms in n-alkanes. Curve 2 shows a gas chromatogram for determining the retention index of substance A, the peak of which is located between two peaks of the n-alkanes n-hexane and h-heptane. Curve 3 shows the determination of the retention index of substance A as the logarithmic interpolation between the retention times of n-hexane and n-heptane. There are four rules for the Kováts retention index:

Rule 1: retention indices for any molecules are composed of the increments for functional groups and of the type of bond. For each CH_2 group, the retention index increases by 100 index units.
Rule 2: with differences in the boiling points of two isomers, the difference between their retention indices on a nonpolar stationary phase is give times larger than the difference in their boiling points.
Rule 3: the retention index of a non-polar substance remains nearly constant for any stationary phase.
Rule 4: functional groups (see also Rule 1) and other compounds as simple compounds enter into the retention index as index increments.

Starting with the index differences, Rohrschneider developed a system for the characterization of separating liquids; the additivity of intermolecular forces was used as a basis. The representation of the retention values is done in the form of retention differences through the correlation of all values with a non-polar standard column (squalane). Benzene, ethanol, butanone, nitromethane and pyridine are used as standard substances which characterize the interactions between the substance and the stationary phase (the separating liquid is characterized by the five Rohrschneider constants).

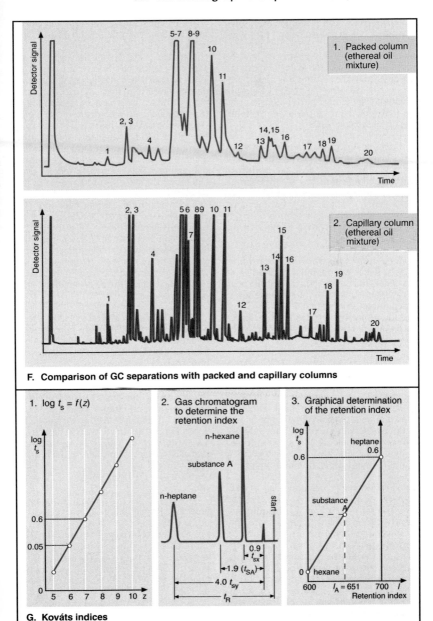

F. Comparison of GC separations with packed and capillary columns

1. Packed column (ethereal oil mixture)

2. Capillary column (ethereal oil mixture)

1. $\log t_s = f(z)$

2. Gas chromatogram to determine the retention index

3. Graphical determination of the retention index

G. Kováts indices

H. Isotherms and temperature-programmed separations. The following correlation exists among the temperature, vapor pressure of the substances and the retention time in GC due to the temperature dependence of the distribution coefficient: the vapor pressure increases logarithmically as the temperature increases, and the retention time is reduced logarithmically at the same time.

1. Isotherm separation. In a homologous series (e.g. of n-alkanes), there is a linear correlation between the peak half-width $b_{0.5}$ and the total retention time t_R: corresponding to the diffusion effects, $b_{0.5}$ increases as the retention time increases. Furthermore, the logarithm of the net retention time t_s is proportional to the carbon number of the homologues.

2. Temperature-programmed separation. With a linear programmed separation of substances in a homologous series, the peak half-widths are constant over a wide range of retentions. Both the total retention time and the respective retention temperatures are proportional to the carbon number of the homologues.

3. Temperature dependence of GC separations. Chromatogram a shows the separation of a four-component mixture with components which vary greatly in their polarity and therefore in their retention temperatures at relatively low temperatures in the isothermal procedure. The analysis time is too long and signal 4 is too wide and is thus relatively low. Chromatogram b shows the same separation at a high temperature: the signals for substances 3 and 4 are now optimally separated in a considerably shorter separation time, but the separation of signals 1 and 2 is insufficient. Finally, chromatogram c shows the results of an optimized temperature-programmed separation which consists of three sub-steps, two isothermal phases at the beginning and end of the separation and a linear temperature increase in the middle.

I. Error recognition in GC analysis.
1. Baselines. The course of the baseline (without sample injection) provides information about possible disturbances: if the detector's measuring sensitivity is low, a straight line parallel to the time axis is to be expected as the baseline or zero line (a). With high detector sensitivities, high-frequency noise (e.g. thermal noise) appears (see p. 158) (b), which can be traced back to the statistical auto-movement of the molecules in the detector, noise from the amplifier and a non-homogeneous temperature distribution in the carrier gas. The noise determines the detection limit of quantitative analyses. Spikes, which extend over and above the short-lived variations in the baseline (the actual noise) (c), can be due to disturbances in the current network. The slow, continual increase (or decrease) of the baseline is called drift (d). Causes of this can include an increase in column bleeding during a temperature-programmed separation or changes in the carrier gas flow. Drifting of the baseline leads to problems with integration (in the determination of the peak areas).

2. Ghost peaks. In contrast to spikes, ghost peaks occur in temperature-programmed separations due to degradation products from the septum (injection component) or to impurities in the carrier gas.

3. Peak asymmetries. Signals having a shape which differs from a Gaussian curve as a symmetrical elution band are called signals with tailing or leading (or fronting), depending on the direction of the deviation. The respective symmetry factor is determined as the ratio of the peak width at a 5% height of the baseline ($b_{0.05}$) and of the segment A in the rising portion of the peak (a).

This can be due to a column being overloaded, e.g. in capillary GC, but it can also occur if the separation temperature is too low (b). A comparison of the signals in the chromatogram shows overloading of the column by substance B. The tailing for a polar substance (nonpolar separating liquid) is due to insufficient inactivation of the carrier material (d).

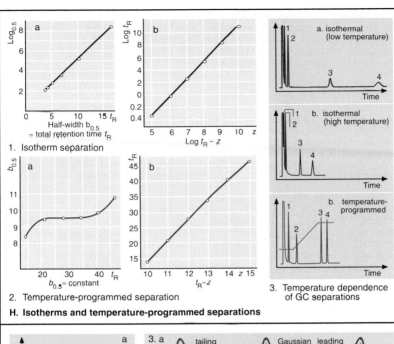

1. Isotherm separation

2. Temperature-programmed separation

3. Temperature dependence of GC separations

H. Isotherms and temperature-programmed separations

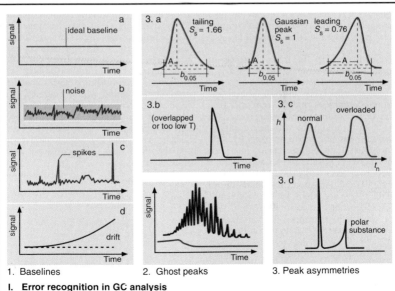

1. Baselines

2. Ghost peaks

3. Peak asymmetries

I. Error recognition in GC analysis

J. Combining separation columns In spite of numerous commercially available separation phases for GC analysis, in day-to-day practice only a few play an important role. One universal separation phase of low polarity is OV-17 (a polysiloxane with phenyl and methyl groups); the separating fluid Carbowax 20M (a polyethylene glycol) is a relatively polar phase. If a mixture of methanol, ethanol and benzene cannot be completely separated either on phase A or phase B—with decidedly different selectivities—a mixture of both separation phases can be used owing to the additivity of the retention times.

K. Deans switch. The serial–parallel switch by Deans makes it possible to avoid transferring the peak to be examined directly to the detector; instead, it is transferred into a second column having different polarity with the aid of a second carrier gas flow. In switch position I, the substances eluted from the non-polar column proceed directly into the detector D_1 (chromatogram a). If one desires to separate off the component x which sits on signal 3 as a tail and to examine it further or to separate it again, then switch II can be used to 'cut out' this part and to redirect it into the polar column, which might have another detector D_2.

L. Gradient trap for trace analysis in gases. An enrichment is often necessary in order to be able to determine trace amounts in gases sensitively enough using gas chromatography. A gradient trap consisting of a piece of column with a casing is used for this. It has a standing temperature gradient: at first, the gas sample from the opposite side is brought via the column casing into contact with very cold air (100 K), and the trace analytes are adsorbed; in the column there is a temperature reduction from about 290 to 170 K. Using hot vapor, the adsorbed substances are transferred in the second step from the gradient trap into a gas sample inlet and then into the GC separation column.

M. Headspace technique. To separate volatile components from the non-volatile or hardly volatile matrix, the headspace technique is used in GC analysis. According to Raoult's and Henry's laws for distribution, the vapor pressure is proportional to the molar fraction, the activity coefficient and the vapor pressure of the pure substance. If a blood or serum sample for the determination of blood alcohol is warmed to a prescribed temperature, then an equilibrium results which can be controlled (using a standard substance such as propan-2-ol) between the liquid and gaseous phases. Using a dosing capillary, a defined volume can be removed from the headspace for GC analysis without the interfering matrix, going over the silicone-rubber membrane of the dosing device and the rubber in the cap of the sample bottle.

N. Heat conductivity detector. The most universally applicable heat conductivity detector uses the differences in the heat conductivities of the carrier gas, e.g. hydrogen with a very high heat conductivity on scale 1, and of the analytes (with considerably lower heat conductivities). The detector (2) consists of a thermostatted metal block with two measuring cells for the reference gas (pure carrier gas) and the eluent flow from the GC column (measuring gas). Heating wires of platinum or tungsten are located in the cells and are attached to a Wheatstone bridge circuit (3). As long as the carrier gas flows through both cells, the heat is dissipated to the carrier gas in the same amount (according to the heat conductivity). If a substance with lower heat conductivity which was separated by gas chromatography enters the measuring cell, this vapor conducts the heat less than the pure carrier gas. Heat is accumulated on the heating wire, and the wire's temperature and therefore also the measurable resistance increase.

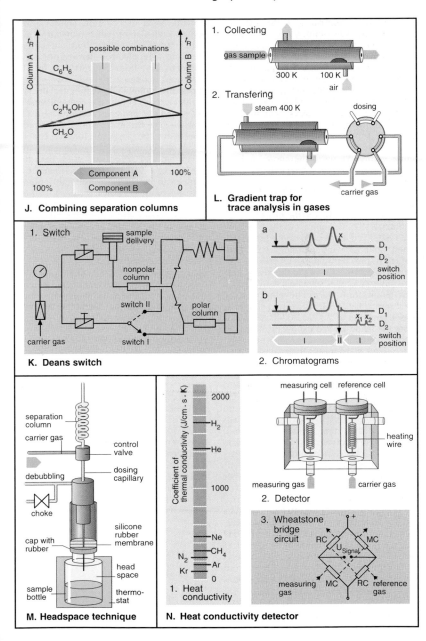

J. Combining separation columns

L. Gradient trap for trace analysis in gases

1. Collecting
2. Transfering

K. Deans switch

1. Switch
2. Chromatograms

M. Headspace technique

N. Heat conductivity detector

1. Heat conductivity
2. Detector
3. Wheatstone bridge circuit

O. Flame ionization detector (FID).

When substances with C—C or C—H bonds are burned in an air–hydrogen flame, ions such as CHO^+ and electrons are formed via radicals. The carrier gas with the substances eluted from the GC column is mixed with hydrogen and air and flows into a burner nozzle, above which a glass piece with the flame tip is located. The flame is ignited using the ignition coil. The ions and electrons form between the anode and the cathode; the current which flows at a particular voltage between the anode and the cathode is recorded as the signal.

P. Thermionic detector (TID).

The TID responds largely to compounds containing phosphorus and nitrogen (therefore it is also called a PND). In contrast to the FID, in a TID there is a glass bead containing rubidium on a glowing platinum wire (salt source heater) between the jet and the collecting electrode. Owing to the low hydrogen flow rate of approximately 3 mL/min, a low-temperature plasma develops instead of a normal flame. The collecting electrode has a potential of 100 to 300 V compared with the flame jet and the alkali metal salt source. With phosphorus- and nitrogen-containing substances, thermal pyrolysis takes place in this flame plasma to form radicals which then react with alkali atoms: $C≡N + Rb = CN^- + Rb^+$. The rubidium cation thus formed is discharged on the negatively charged bead and the cyanide ions migrate either to the collecting electrode, or they are further combusted, releasing an electron. This detector can be used to detect pesticides such as phosphates and N-methylcarbamates with high sensitivity and selectivity compared with other organic compounds.

Q. Electron capture detector (ECD).

As a selective detector (1) for halogen, sulfur, heavy metal and nitro compounds, the ECD has a β-emitter (such as tritium–copper foil or ^{63}Ni) which ionizes the carrier gas (e.g. helium), whereby slow electrons are released. These primary electrons have only low (thermal) energy and are led to the anode due to the applied acceleration potential (zero point). If the carrier gas flowing from the column contains substances with electron-absorbing properties, the zero point is reduced by the uptake of electrons. The electron density present in the detector, the zero point, is usually measured using a pulsed d.c. voltage applied to the anode. Both the pulse interval (3) and the level of the d.c. voltage must be optimized for the substances to be determined. The examples show the dependences for the pesticide lindane and for tetrabromoethane.

Helium and argon ionization detectors work using a similar principle. However, in contrast to the ECD, considerably higher voltages of up to 1000 V are used. In these ionization detectors, the helium or argon is present principally in an excited state; substances with a lower ionization potential than the excitation energy of the metastable helium or argon are ionized. Compounds such as methane, nitrogen and oxygen cannot be detected in an argon ionization detector owing to the excitation potential of only 11.7 eV (compared with 20.6 eV for helium). Thus, the ionization detectors differ in their selectivity.

R. Comparison of different types of detectors.

1. HCD and FID. The aqueous solution of an apple aroma was separated on a polyethylene glycol capillary column using temperature programming from 70 to 230 °C. The organic substances under the water peak can also be detected by FID.

2. FID and ECD. The differences in detection between FID and ECD are made clear using chlorinated phenols as an example: all polychlorinated phenols are detected more sensitively with the ECD than with the FID (3–5-fold, e.g. 2,4,6-trichlorophenol). On the other hand, 4-chlorophenol is not detected by the ECD.

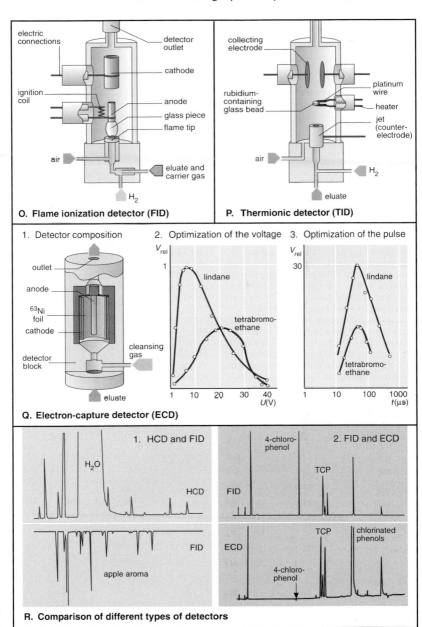

O. Flame ionization detector (FID)

P. Thermionic detector (TID)

1. Detector composition
2. Optimization of the voltage
3. Optimization of the pulse

Q. Electron-capture detector (ECD)

1. HCD and FID
2. FID and ECD

R. Comparison of different types of detectors

9.3 Electrophoresis

A. Principles of electrophoretic separation. Separation methods which are based on different migration rates of electrically charged particles in an electrolyte solution under the influence of an electric field are grouped under the term electrophoresis. There are essentially three different principles of electrophoretic separation. Zone electrophoresis (with a carrier or carrier free) uses a homogeneous buffer system in which the pH is constant over the entire separation path and separation time. The migration distances traveled in a prescribed time correspond to the electrophoretic mobilities m_{RA} and m_{RB}.

Isotachoporesis uses a discontinuous buffer system in which the ions which have separated owing to differing mobilities each move between two electrolytes, the faster leading ion L^- and the slower terminating ion T^-. The ions with the highest mobility follow directly after the leading ion; those with the lowest mobility migrate directly in front of the terminating ion.

Isoelectric focusing is used exclusively for amphoteric substances such as proteins with differing isoelectric points. In a stable pH gradient, they migrate to the location with the pH value of their isoelectric point, pI, in the direction of the anode or cathode, depending on their charge.

B. Principles of zone electrophoresis. Starting with the ion mobility in infinitely dilute solutions, the electrophoretic mobility of a particle is defined for a field strength of 1 V/cm using Stokes' law of friction. This factor is used for the characterization of the individual substances (such as retention in chromatography). In zone electrophoresis, the same pH and field strength re present throughout the entire separation area. The sample solution is applied in dots or lines to a carrier (e.g. a paper strip) soaked in the electrolyte. The electrolyte takes over the function of the current transporter here. After the electric field has been applied, the ions migrate—anions A^-, B^- and C^- in this case—independently of the electrolyte ions, based on their electrophoretic mobility.

C. Principles of carrier-free electrophoresis. Carrier-free electrophoresis in a U-shaped tube is also called the Tiselius method or the method of migrating interfaces. The sample solution with the three anions is overlaid in both arms of the U-tube with a pure buffer solution; the latter solution must be specifically lighter than the sample solution in order to prevent convectional currents which would destroy the interfaces. After electric current has been applied, there is partial demixing. The anions with the greatest mobility are near the anode and those with the least mobility are after the sample solution, in the direction of the cathode. However, complete separations cannot be accomplished using this principle.

D. Techniques for carrier-free and carrier electrophoresis. Carrier-free electrophoresis uses platinum electrodes in a U-tube; the electrical field is generated by a d.c. source. Migrating fronts (see C) develop as a result of the electrophoretic migration in the buffer solution. This technique is also called interface electrophoresis.

In carrier electrophoresis, the electrophoretic mobilities of ions are smaller than in free electrophoresis (owing to the channel effect of the carrier made of paper or a gel and to electroosmosis). For example, in paper electrophoresis, filter-paper is soaked in the electrolyte solution. A voltage of $100+$ V is applied to the strip of filter-paper, and the mixture is applied at a particular spot on the strip of paper. The carrier's structure contributes to the electrophoretic separation. After a sufficiently long separation time, the zones of the individual ions are separated from one another by empty zones (with just electrolyte ions).

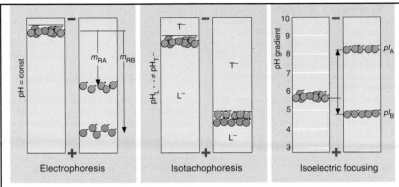

A. Principles of electrophoretic separation

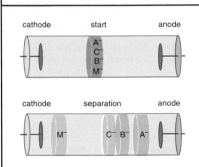

B. Principles of zone electrophoresis

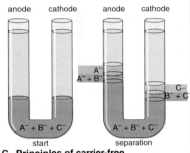

C. Principles of carrier-free electrophoresis

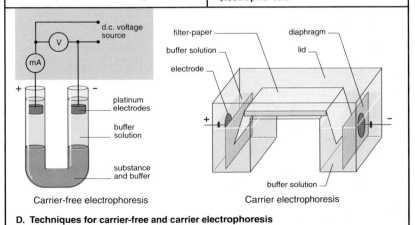

D. Techniques for carrier-free and carrier electrophoresis

E. Techniques of carrier electrophoresis. *1. Column electrophoresis.* Column methods are particularly well suited for preparative separations, e.g. as shown here with a polyacrylamide gel. The electrodes consist of stainless steel, platinum or carbon, and as in carrier electrophoresis, they are made of thin layers of cellulose, cellulose acetate, agar, agarose or starch gel and polyacrylamide gel on insulating plates. The electrode areas are separated by diaphragms of the buffer compartments. The resultant current heat is rerouted by inner and outer cooling tubes in column electrophoresis. The migration rates in a column are less than 1 mm/min; it is possible to reduce the length of the electrophoresis using high-voltage electrophoresis (up to 10000 V). After the power has been turned off, the substances separated in the column can be recovered by fractionation.

2. Continuous carrier electrophoresis. In the continuous procedure, buffer solution and the sample are added continuously from above to a carrier layer arranged vertically, which can be made of paper. Owing to the effects of gravity, both solutions flow toward the bottom. A low-voltage field (e.g. 300–400 V) applied horizontally diverts the different sample ions in paths which correspond to their electrophoretic mobility. The fractions obtained in this way can be captured in separate vials via the serrated, protruding end of the carrier.

F. Applications of carrier electrophoresis. *1. Bridge for cellulose acetate strips.* Cellulose acetate strips are often used as a carrier material in carrier electrophoresis. They are stretched in strips across a bridge which is placed in an electrophoresis chamber. The strips dip directly into the electrode buffer. Analytical separations can be achieved with this technique.

2. Electrophoresis of isoenzymes. Electrophoresis with cellulose acetate is used above all in clinical chemistry, for example in the diagnosis of particular diseases. One can obtain electrophoretic separations with proteins similarly to thin-layer chromatography, by dyeing in the form of electropherograms. The example shows the densitometric analysis (see also p. 152) of serum lactate dehydrogenases after separation in a sodium veronal buffer and their assignment to specific diseases.

3. Serum electropherogram. Albumins and globulins in serum can be separated by paper electrophoresis in a similar manner. A serum electropherogram is shown after being dyed with a blue dye; below this is the photometric analysis in thin-layer chromatography using a scanner (see p. 152). Diagnoses of particular diseases are possible here too by making comparisons. The sharpness of separation can be optimized by the selection of the buffer system (especially by the ionic strength).

G. Flow-through electrophoresis. Column electrophoresis (see E.1.) can also be evaluated analytically, as shown by the example of the separation of soya proteins. Using an elution buffer, the electrophoretically separated proteins are pumped through an elution chamber at the lower end of the separating gel and then into the measuring cell of a UV detector. A characteristic distribution pattern is seen at 280 nm; however, separation times of more than 10 h are required with a gel length of 4 cm and a flow rate of 0.05 mL/min for the elution buffer. The optimal preparation of the gel (careful pouring of the gel)—polyacrylamide gel here—is one of the most important prerequisites for electrophoretic separations which are to be evaluated analytically. Similarly to gel chromatography (see p. 146), but using a different separation principle, the electropherograms can be evaluated in order to characterize protein mixtures.

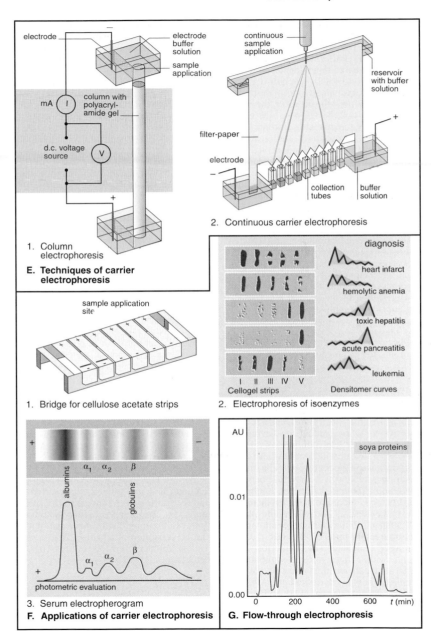

1. Column electrophoresis

2. Continuous carrier electrophoresis

E. Techniques of carrier electrophoresis

1. Bridge for cellulose acetate strips

2. Electrophoresis of isoenzymes

3. Serum electropherogram

F. Applications of carrier electrophoresis

G. Flow-through electrophoresis

H. Disk electrophoresis. The special technique of disk electrophoresis shows a discontinuity above and beyond the use of a discontinuous buffer (see isotachophoresis). Zone sharpening is achieved due to the pH of the buffer and especially to the gel structure and the ionic strength of the buffer. Thus, in addition to the pH value, the buffer composition and the gel pore size of the separating system are also discontinuous. Glycinate can be used as the trailing ion; its net mobility is highly dependent on the pH value. There is a higher pH in the separating gel than in the part where the sample solution is brought at the start of the separation (sample or collecting gel), whereby the trailing ions achieve a higher motility than the ions in the sample. Chloride ions with their high mobility are used as leading ions. The concentrated sample ions are passed by the trailing ions while still in the collecting gel. The zone sharpening in this technique results from the effects of the use of leading ions and trailing ions, the pH dependence of the migration of the trailing ions and the influence of porous gels against interference due to convection (1).

Compared with carrier electrophoresis (F.3), with disc electrophoresis of serum proteins one can obtain a better separation of the β-globulins (2).

I. Isotachophoresis. *1. Schematic diagram of the principles of anion separations.* The discontinuous electrolyte system consists of a leading electrolyte L and a trailing electrolyte T, the leading ion having the greater mobility. The mobilities of the ions to be separated must lie between those of the leading ion and the trailing ion. The sample solution is applied at the interface between the leading ion and the trailing ion: in the start phase L^- ions are in the anode region and in the capillary (the separation system), T^- ions are in the cathode region, and in between are the ions A^-, B^- and C^- which are to be separated. The counterion is the same within the entire electrophoretic separation path and it has the effect of buffering to provide the desired pH. After the electric field has been applied, each type of ion arranges itself between the leading and trailing ions, resulting in discrete zones.

2. Course of mobility, field strength and concentration. After the electric field has been applied, the weakest field strength is in the area around the leading electrolyte, while the highest field strength is around the trailing electrolyte. The stationary state of the separation is characterized by a step-like course of the field strength, the mobility and the concentration. The field strength increases stepwise along the zonal interfaces. The field strength and concentrations are constant in the discrete zones.

3. Effect of zone sharpening. The potential jumps necessitate a constant migration rate of all ion types; the product of the field strength and the net ion mobility is constant. The zone formation causes the slow trailing ions to be retarded by the lower field strength of the zone in front of them; faster leading ions are accelerated by the higher field strength of the following zones, and a dynamically induced zone sharpening occurs as a result. At a prescribed leading ion concentration, at a stationary state, the concentrations of all ions adjust themselves so that zones with previously low concentrations are concentrated and those with higher concentrations are diluted. Therefore, the width of the zone represents a measure of the amount of a type of ion.

4. Schematic diagram of an isotachophoresis device. The separation system consists of a Teflon capillary of perhaps 0.5 mm inner diameter and 20 cm in length, which is thermostatted. Leading and trailing electrolytes are injected separately. A UV detector can be used for detection.

5. Separation of serum proteins. Using the leading ion morpholinoethanesulfonate and the trailing ion aminocaproate (pH 9 and 10.8, respectively), 16 individual fractions of immunoglobulins in human serum can be separated in less than 25 min in a Teflon capillary 23 cm long followed by UV detection.

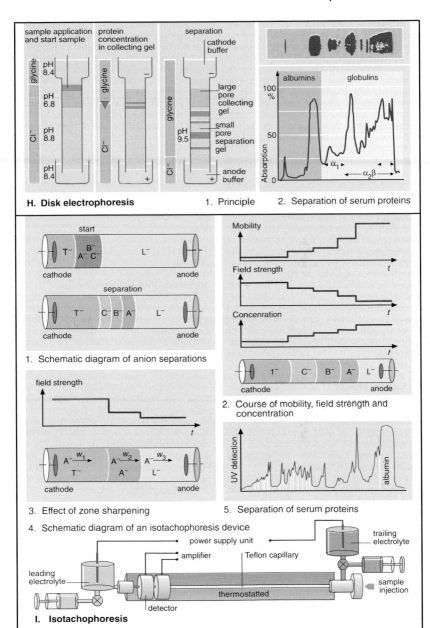

H. Disk electrophoresis 1. Principle 2. Separation of serum proteins

1. Schematic diagram of anion separations

2. Course of mobility, field strength and concentration

3. Effect of zone sharpening

4. Schematic diagram of an isotachophoresis device

5. Separation of serum proteins

I. Isotachophoresis

J. Isoelectric focusing. *1. Principles.*
Amphoteric molecules migrate in an electric field within a pH gradient into the zone of their isoelectric point. The p*I* value is defined as the pH at which ampholytes are present as zwitterions, without a net charge. At higher pH, ampholytes migrate in an electric field as ions; at lower pH, they migrate as cations. The pH or buffer gradient is set up so that the area of the lowest pH is located at the anode, while that of the highest pH is located at the cathode. After the current has been applied, the separation of substances with different isoelectric points is complete when all of the ampholytes are in the same pH region as their p*I* value, and are therefore in a stationary state. This procedure is called isoelectric focusing. The necessary stationary pH gradients can be produced using density gradients of electroneutral substances in which locally different buffers (with different pH values) are dissolved. Carrier ampholytes (polyaminocarboxylic acids) are used for these buffer zones; they establish pH gradients which are stable even in an electric field as a result of a stationary juxtaposition. Under optimal conditions, separations can be attained even with differences in isoelectric point of 0.01 to 0.02 pH unit.

2. Analysis of the origin of animal proteins.
In food analysis, proof of the type of animal or plant from which a foodstuff is derived often plays an important role with respect to counterfeiting (e.g. a roast from a domestic pig instead of a wild boar). Isoelectric focusing can often be used to detect species-specific protein zones in electropherograms of the sarcoplasm (muscle) proteins. This method makes it possible to differentiate between cattle, swine, horses and sheep and the wild types. To perform the analysis in the pH range 3.5 to 9.5, the proteins are extracted in the presence of sodium dodecylsulfate (SDS) to increase their solubility. The proteins are converted from their native condition into anionic SDS protein micelles. SDS electrophoresis on starch, agarose or polyacrylamide gels is the standard method of electrophoretic protein separation.

K. Immunoelectrophoresis. *1. Three different techniques.* Immunoelectrophoresis combines an electrophoretic separation with an immunological reaction (identification, see p. 48). The principle behind immunoelectrophoresis is the formation of precipitation lines at the equivalence point between antigen and antibody. The immunoprecipitates are visible as a white line in a gel. They can also be colored with protein dyes.

In countercurrent electrophoresis (a) in an agarose gel, the antibodies do not have a charge at pH 8.6. The sample and antibodies are pipetted into the appropriate openings, and they migrate toward one another—the charged antigens due to electrophoresis and the antibodies due to the electroosmotic flow.

In zone electrophoresis with immunodiffusion (agarose gel), after electrophoresis there is a diffusion of the antigen fraction against the antibodies which are pipetted into the side grooves (b).

In the rocket technique (c), the antigens migrate electrophoretically into an agarose gel with the antibodies. The areas of the rocket-shaped precipitation lines are proportional to the concentrations in the samples.

2. Course of immunoelectrophoresis. The sample mixture is applied to a gel (a), then there is an electrophoretic separation using zone electrophoresis (b). When the electrophoresis is concluded, a groove is cut out along the separation path and filled with antiserum (antibodies). As a result of the diffusion into the gel, precipitation lines develop at those points where immune complexes form between antibodies and antigens.

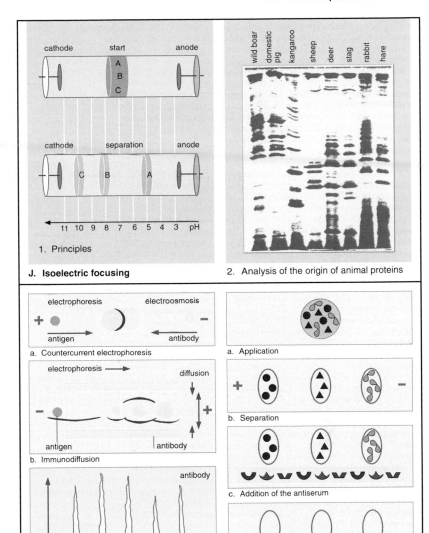

cathode   start   anode

A
B
C

cathode   separation   anode

C   B   A

11 10 9 8 7 6 5 4 3 pH

1. Principles

J. Isoelectric focusing

2. Analysis of the origin of animal proteins

wild boar | domestic pig | kangaroo | sheep | deer | stag | rabbit | hare

electrophoresis   electroosmosis

+   −

antigen   antibody

a. Countercurrent electrophoresis

electrophoresis →   diffusion

−   +

antigen   antibody

b. Immunodiffusion

antibody

antigens

c. Rocket technique

1. Three different techniques

K. Immunoelectrophoresis

a. Application

+   −

b. Separation

c. Addition of the antiserum

d. Precipitation

2. Course of immunoelectrophoresis

L. Capillary electrophoresis. Capillary electrophoresis (high-performance capillary electrophoresis, HPCE) represents the newest technique for electrophoretic procedures (first described in 1979). In principle, all of the electrophoresis methods which have been described can be performed today in an apparatus with thin glass capillaries, viz. fused-silica capillaries as described in capillary GC (see p. 164) with inner diameters of 25 to 100 μm and lengths of 20 to 100 cm.

1. Schematic diagram of a capillary electrophoresis apparatus. Each end of a capillary filled with buffer is immersed in a separate reservoir of buffer. The electrodes from the high-voltage supply are also immersed in these buffer solutions. The power source can be set at up to about 30 kV at a current of up to 100 μA. The detection system, e.g. a UV detector, is positioned just before the outlet of the separating capillary. A special system of lenses is used to focus light on the capillary with wavelengths between perhaps 190 and 380 nm. All components are largely miniaturized: volumes of the electrode chambers, about 1 mL; that of the inlet block, about 0.1 mL; diameter of the optics, approximately 2 mm. The detection system produces a signal proportional to the amount or concentration of the substance, and the signal can be passed to a computer system. In addition to the capillary (see 2), the inlet block with the sample application device is an integral component of this technique: owing to the small total volume in the capillary (only about 1 μL with a length of 50 cm and a 50 μm inner diameter), the sample cannot contain more than 1–10 nL in order to ensure narrow start zones. The principle behind all sample application systems for capillary electrophoresis lies in the fact that larger volumes in the μL range are present, but only nL volumes of this can be transferred to the capillary. For gravity injection, a difference in level is produced by raising the sample vessel (or lowering the buffer vessel at the outlet block): a hydrostatic flow results, based on the principle of communicating tubes, and the differences in height and time determine the volume flowing into the capillary.

In electrokinetic applications, the sample migrates into the capillary electrophoretically by applying a strong electric field.

In vacuum or pressure applications, the pressure difference and the time determine the applied volume.

2. Cross-section of a quartz capillary. The capillary should be housed in a thermostatted cassette, which is made of quartz in many cases in GC (see also p. 164). The quartz capillaries with inner diameters between 10 and 200 μm are covered with polyamide. Thick-walled capillaries have the advantage that they divert the Joule heat which develops more efficiently at higher voltages owing to the increased heat capacity. The UV transparency of quartz plays an important role in UV detection. Undesirable adsorption effects on the walls of the capillaries can be avoided by deactivating the inner walls by coating them with polyacrylamide. It is also possible to pack using a gel material.

3. DNA analysis. The major areas of application for capillary electrophoresis are protein and DNA (deoxyribonucleic acid) analysis. In a 35 cm × 50 μm capillary, DNA fragments with 88 to 1746 base pairs can be separated within 10 min using a borate buffer with 8 kV voltage and detection at 260 nm (a). Conventional polyamide gel electrophoresis would require 3 h for this (b).

4. Separation of reducing sugars. Capillary electrophoresis can also be used successfully to separate groups such as reducing sugars, e.g. as anionic *N*-2-pyridylglycosamine derivatives in the example —using 15.0 kV, borate buffer of pH 10.5 and UV detection at 240 nm.

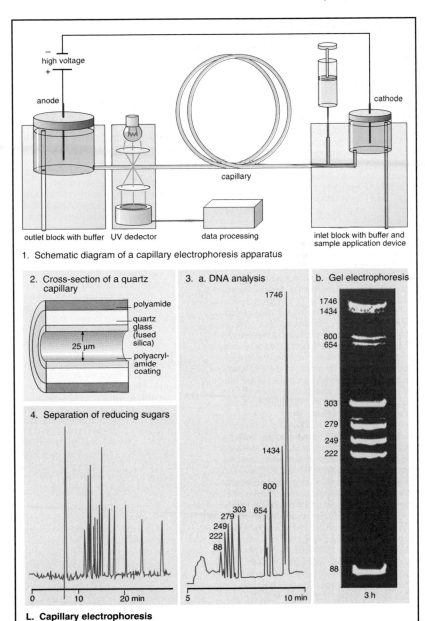

1. Schematic diagram of a capillary electrophoresis apparatus

2. Cross-section of a quartz capillary

polyamide
quartz glass (fused silica)
25 μm
polyacrylamide coating

3. a. DNA analysis

b. Gel electrophoresis

4. Separation of reducing sugars

L. Capillary electrophoresis

10 Automation of Analytical Procedures

Introduction. Large numbers of samples or of steps in an analytical procedure often make it necessary to automate or partially automate the process, especially in routine analysis. In industry in the Federal Republic of Germany alone, in laboratories for clinical analysis, environmental analysis and production monitoring, it is estimated that several million samples are analyzed annually. Especially in the last 10 years, time and cost reductions have led to increased interest in the possibility of automation.

For chemical-analytical procedures, automation was first used in clinical-chemical analysis—represented by the principle of continuous flow analysis (CFA) (see p. 186) with air-segmented liquid streams (since 1957) and its further development, flow injection analysis (FIA) (since 1975, see p. 190). These systems are also called chemical analysis machines (or Auto-Analyzers in the field of clinical chemistry). The next step in the automation of analytical procedures is characterized by the use of analytical robots or robot systems in connection with powerful computers since the mid-1980s. Particularly the advances in sensor and microprocessor technology have led to this development.

A. Flow chart for determining water- and citrate-soluble phosphate. The starting points for automating operational steps in an analytical laboratory are obtained by modularly breaking down an analytical protocol into basic operations. Each of the basic operations should be performed by a machine, with which the assigned parameters of the respective analytical step can be programmed. The analysis of a fertilizer is illustrated here using the symbols shown on p. 9 as per DIN 32650 and 32649. The equipment needed is shown to the left of each symbol, and the required reagents or other details about the steps are shown on the right.

From this figure, the idea of an automatic analytical system can be conceived from the following elements: the basic operation machine (see B), the sample transport (e.g. in flow systems, see 10.1), the central control, the input station and the output of the results.

The basic operations include weighing, dosing, dissolution, dilution, liquid–liquid and solid–liquid extraction (including phase separation), titration and photometry. Basic operation machines execute the named processes under the control of a microprocessor; the data are transmitted to the machine by the central control.

The input station, one of the two interfaces between the entire system and the user, includes the scale, the identification of the analytical sample and the introduction of the sample into the sample transport area.

The output of results is the second interface between the system and the user. It also proceeds via central control through computer systems. The central memory contains all of the information about the total analytical process and the protocol data from each step for each sample analyzed, in addition to the necessary calculation programs.

B. Workstation functions. 'Liquid manipulation' is a function at the core of a workstation. For example, 12-way valves make it possible to direct liquid flows to and from the different functions such as reservoirs, dosage devices, membrane filtration and solid-phase extraction (see also p. 30). These dosage steps are usually monitored on a scale and documented in data files. Laboratory robots go the next step; based on their mobility (using servodrives), they can take over tasks in sample preparation.

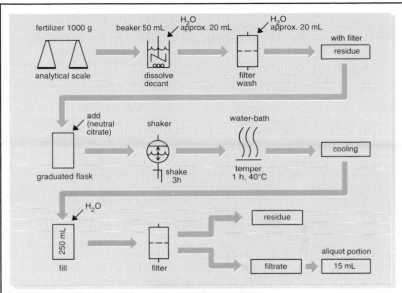

A. Flow chart for determining water- and citrate-soluble phosphate

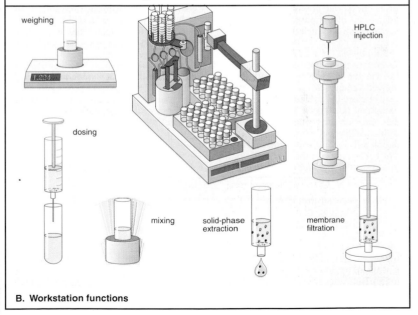

B. Workstation functions

10.1 Continuous Flow Analysis (CFA)

Continuous flow analysis was introduced in 1957 by Skeggs for photometric analysis with high sample numbers in clinical chemistry.

A. Structure of the analytical system.
1. Flow Chart of CFA. A peristaltic pump transports the solutions (reagents). The liquid flow is segmented by the introduction of air (in the form of air bubbles). The samples to be analyzed are likewise brought into the reaction system, a spiral-shaped, wound glass tube (mixing coil) using a hose (sample hose). A defined sample volume is maintained by dipping the sample tube into the sample vessel in conjunction with an automatic sample injector for only a specified time, between 10 and 60 s. During the intervening intervals (approximately 30 to 60 s), it takes up distilled water or a buffer solution. Mixing, and therefore the chemical reaction, take place in the mixing coil. The reaction time is determined by the length of the mixing coil (of the glass tube) and also by the flow rate of the liquid flow. Before the liquid flow reaches the flow-through cell of the photometer, the air is removed again by a debubbler (see 2).

2. Components of a CFA system. The air required for the segmentation of the liquid flow via the debubbler is quantitatively removed (along with a small volume of liquid) before the measuring cuvette. The mixing coil is connected with the reagent tubes and with the air and sample tube by 'cacti'. Additional components are the mixing coils (see 1) and standardized and color-coded tubes for the peristaltic pump; these transport the defined volume as a factor of the pumping speed.

B. Air-segmented liquid flow.
The purpose of segmenting the air is to keep the diffusion effects low in the longitudinal direction, and also to prevent or minimize diffusion resulting from differences in the flow rates in the middle of the flow and along the walls of the glass tube and hoses (flow profile). Both effects would result in mixing together of sequentially applied samples along the way to the photometer. However, even with air segmentation, a liquid film with thickness d_t forms on the walls. In this way, liquid from one segment arrives in several other segments, leading to dispersion (band broadening) of the sample analyte.

C. Dispersion dependences.
Dispersion, shown as the band broadening s of a Gaussian curve as in chromatography (see p. 144), depends on the following physical-chemical values in the form shown: the surface tension of the liquid flow γ (1), the diameter of the mixing coil tube and of the hoses d_t (2), the length of the total flow system l (3), the linear flow rate U (4) and the viscosity of the liquid flow η (5).

D. Signal forms.
The flow systems we have described make possible exact volume dosing (of the sample and the reagents) and exact timing of the individual reaction steps: even reactions which do not proceed completely due to the time prescribed by the flow system can be used in a CFA system. If one allows a sample with a constant concentration to flow through the system for a prolonged period, a maximum (plateau) appears, which indicates that the reaction is complete: one obtains a steady-state curve (1). In actual practice, shorter suction times (of the sample solution and water or buffer) are selected, so that signals without a plateau (on a baseline = reagent blank value) are obtained which are reproducible and the height of which is dependent on the sample concentration (2).

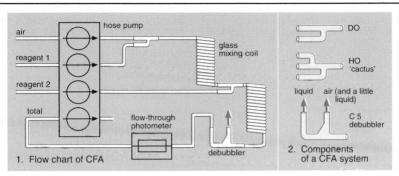

A. Structure of the analytical system

1. Flow chart of CFA

air
hose pump
reagent 1
glass mixing coil
reagent 2
flow-through photometer
total
debubbler

2. Components of a CFA system

DO

HO 'cactus'

liquid   air (and a little liquid)

C 5 debubbler

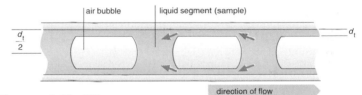

air bubble   liquid segment (sample)

$\frac{d_t}{2}$ d_t

direction of flow

B. Air-segmented liquid flow

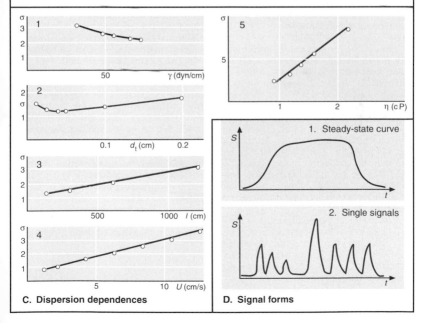

C. Dispersion dependences

1   γ (dyn/cm)   50
2   d_t (cm)   0.1   0.2
3   l (cm)   500   1000
4   U (cm/s)   5   10
5   η (cP)   1   2

D. Signal forms

1. Steady-state curve

2. Single signals

E. Automatic fluorimetric catecholamine determination. The catecholamines adrenaline and noradrenaline belong to the biogenic amines which are formed during physiological reactions in organisms. Information about the amounts of these amines in blood and urine is important for physicians in connection with stress examinations and for diagnostic purposes, especially in tumor diagnosis.

1. Conversion to fluorescent derivatives. Adrenaline and noradrenaline can be determined particularly sensitively and selectively after their oxidation to trihydroxyindoles. First they are oxidized by potassium hexacyanoferrate(III) via *o*-quinones with cyclization to give adrenochromes. In the second reaction step, isomerization to form trihydroxyindole derivatives takes place in an alkaline environment with the addition of a reducing agent. The fluorophores are stabilized by lowering the pH of the solution to about 5. All of these intermediate steps must be executed in time intervals which must be exactly maintained. Continuous flow analysis is especially well suited for this requirement of exact timing.

2. Flow chart for adrenaline/noradrenaline determination. The numbers on the pump hoses correspond to the following solutions: 1 = buffer, 2 = sample solution, 3 = air, 4 = oxidation solution, 5 = only with the separate supply of reduction solution (sodium sulfite and mercaptoethanol) and sodium hydroxide solution, 6 = NaOH–reducing agent, 7 = acetic acid and 8 = total tube. The letters DO, D1 and HO indicate connecting pieces (see C.2) and C5 is the debubbler. The system contains two mixing coils (MC and DMC = double mixing coil).

3. Fluorescence spectra. There is so little difference in the fluorescence spectra of the two trihydroxyindoles at an excitation wavelength of 400 nm that a differentiated or separate analysis of the two amines is not possible with fluorescence spectroscopy.

4. Selectivity effects via the speed number. The speed number is the pumping rate, from which the flow rate (with defined tube diameters) and the reaction times (with established mixing coils) for the individual partial reaction steps are derived. Owing to the different kinetics of the aforementioned reactions, these device parameters can be used to determine adrenaline selectively in addition to noradrenaline (increase in speed number = reduction in reaction time). If one also includes the pH dependence of the oxidation, the following options become available: at pH 6.0 (a), noradrenaline is oxidized faster than adrenaline; the reverse occurs at pH 2.8 (b).

5. Measuring signals for determining adrenaline in addition to noradrenaline. At a speed number of 10 and oxidation at pH 2.8, low concentrations of adrenaline—as a stress parameter (10 ng/mL here)—can be selectively determined in an excess of noradrenaline (100 ng/mL) (excitation wavelength 400 nm, fluorescence wavelength 490 nm, reaction temperature 32 °C).

F. Cyanide determination. Even complicated analytical procedures requiring additional steps besides the exact dosing of reagents in intervals which must be exactly adhered to can be performed with continuous flow analysis systems: the flow chart for cyanide determination shows a distillation step for separating free cyanide from effluent samples. After acidification with phosphoric acid, the distillation is conducted using a bath temperature of 140 °C, then the solution is cooled and gas bubbles which develop are removed using the debubbler before the actual chemical reaction with chloramine-T and pyridine and barbituric acid takes place to form a polymethine dye (at a 40 °C reaction temperature). Very low photometric detection limits in the lower micrograms per liter region can be achieved using 50 mm cuvettes.

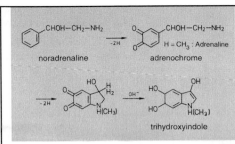

noradrenaline   adrenochrome

H = CH₃ : Adrenaline

trihydroxyindole

1. Conversion to fluorescent derivatives

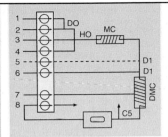

2. Flow chart for adrenaline/
   noradrenaline determination

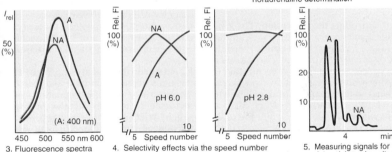

3. Fluorescence spectra

4. Selectivity effects via the speed number

5. Measuring signals for
   determining adrenaline
   and noradrenaline

E. Automatic fluorimetric catecholamine determination

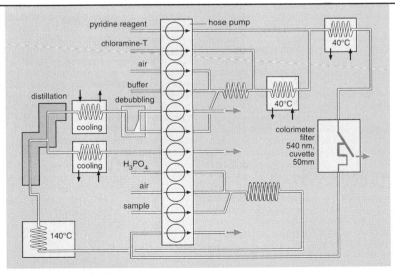

F. Cyanide determination

10.2 Flow Injection Analysis (FIA)

A. Principles of the FIA technique. A prescribed, reproducible sample volume is injected into an unsegmented continuously flowing liquid (carrier) stream, e.g. into a reagent solution. The sample is entrained by the reagent flow; there is a mixing of sample and reagent solution via convection and diffusion. However, this mixing, a prerequisite for the conversion of the analyte into a detectable reaction product, is not homogeneous and does not necessarily have to reach an equilibrium.

B. Controlling the dispersion. The schematic depiction of a dispersed sample zone produces a concentration profile, whereby the dispersion D, which occurs as a result of the sample segment with concentration C_0 mixing with the reagent flow, can be calculated as the change in peak height with $D = C_0/C_{max}$. Therefore, D is also an approximate measure of the dilution of the sample in the FIA system. The course of the curve also shows that under the instrumental conditions, a defined concentration in the dispersion zone can be assigned to each point in time. Owing to the reproducible timing, one obtains controlled dispersion which can also be used analytically. For example, if the linear region of a measuring method is surpassed at the maximum with C_{max}, any other point can be used for evaluation–a process of a gradient dilution.

C. Characteristic signal forms. The concentration profiles which one can record with the FIA technique after the injection of a dye depend on the sample's residence time in the system. Laminar flow conditions rule during transport in the narrow tubes (0.3 to 0.8 mm inner diameter). If a sample is brought into the system, after some time under zero-flow conditions it will have dispersed itself to form something close to a Gaussian curve. There is an axial dispersion of the sample zone as a result of convection processes; in practice, radial diffusion also occurs. The molecules at the tip of the sample zone move outward (in zones of low flow rate), while those at the end of the zone diffuse into regions of higher flow rate. This produces the signals shown with different bandwidths, which increase with the length of the transport path.

D. Dependence of the signal forms on the sample volume. The selection of the sample volume has a considerable and effective impact on the dispersion. With a high sample volume of 800 μL, there is no mixing; in this case, the sample zone is not affected by intermixing with the surrounding transport flow. One speaks in this case of limited dispersion, and the only task of the FIA system is that of a transport system. On the other hand, if one or more reagents need to be mixed together for photometric analyses, one selects a dispersion of 3 to 10 to achieve a sufficient level of mixing, i.e. reaction.

E. Stopped flow analysis. With slow reactions, the reaction time and therefore the sensitivity of the determination can be increased without any increase in dispersion by briefly stopping the pump—called the stopped-flow technique. In this example of enzymatic glucose determination, the stop time was 15 s in each case.

F. FIA titration. If one injects an acidic sample into the carrier stream (NaOH with bromothymol blue as an indicator here), the concentration gradients (see B) which form on either side of the sample zone can be compared with the course of a titration curve. There is an acid excess at the maximum; an equivalence point is reached in each of the ascending and descending branches, with the same dispersion. The concentration of the injected acid is derived from the time difference. The transition of the indicator from blue (alkaline) to yellow (acid) is recorded photometrically.

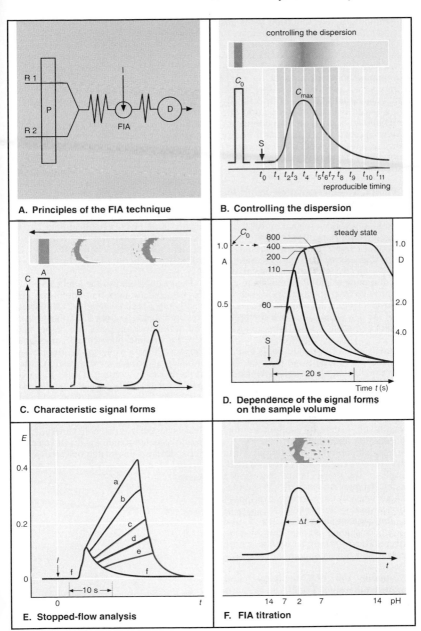

A. Principles of the FIA technique

B. Controlling the dispersion

C. Characteristic signal forms

D. Dependence of the signal forms on the sample volume

E. Stopped-flow analysis

F. FIA titration

G. Ammonium determination using gas diffusion. As with the CFA technique, in FIA flow systems other analytical steps can be integrated as separation techniques, as shown here with gas diffusion. To determine ammonium levels in water, ammonia is transported through a gas-permeable membrane (usually PTFE, or specifically Teflon) via diffusion after alkalization with NaOH into an acceptor stream containing an indicator solution. The indicator's color change as a result of the acid–base equilibrium which develops is used for photometric evaluation.

H. Phosphate determination using the FIA technique. *1. Flow chart.* For the quantitative analysis of orthophosphate ions, solutions with molybdate and ascorbic acid are mixed in a chemifold consisting of a PTFE tube 52 cm long and having an internal diameter of 0.75 mm. The sample volume is injected into the injection valve S. The conversion to molybdophosphate and then to molybdenum blue occurs in a second chemifold 50 cm long, and the concentration of the dye is measured in a flow photometer as absorption at 660 nm.

2. Signals. In this phosphate analysis procedure, 30 μL of sample are injected. The residence time in the second chemifold from the point of injection S_1 until the peak maximum R is reached in about 15 s. An additional 15 s are necessary to rinse the sample zone out of the measuring (flow-through) cell, except for a slight residual concentration until the next injection S_2. The signals represent temporally spread out curves for concentrations of 20 or 40 μg/L phosphate. The signals before that show a possible sample frequency of 120 samples per hour, whereby there are only overlaps of two successive sample zones in the range of 1% for concentrations between 5 and 40 μ/L. Depending on the flow system, between 60 and 360 samples per hour can be analyzed with this technique.

I. Coupling the FIA technique and microwave digestion. The FIA technique is also suitable for coupling with digestion methods (see p. 24). To determine the total phosphorus content in effluent samples, the possible presence of condensed phosphates and organic phosphonic acids makes it necessary to digest or convert all phosphorus compounds into orthophosphate ions. In the flow chart shown here (1), a decomposing agent consisting of perchloric acid and peroxodisulfate is mixed with a sample of 100 μL in an FIA system. The hydrolysis of triphosphate or the oxidation of phosphonic acids to orthophosphate is done in a microwave oven. Gas bubbles which form (especially due to the oxidation of organic substances to carbon dioxide) are removed from the FIA system by a gas diffusion cell. A molybdate solution is mixed in as reagent R_2 and the molybdophosphate which is formed is detected amperometrically in an electrochemical detector, as used in HPLC analysis. The reduction is accomplished electrochemically (ELCD) and not with a chemical reducing agent as in H.

The example shows the applicability of this detector instead of a photometer with readily reproducible signals (2) and detection limits below 0.1 mg/L of phosphorus. The decomposition ratios are higher than 90% for organic phosphorus compounds and about 70% for triphosphate. Besides gas diffusion and microwave digestion, other methods can be combined with FIA systems, such as dialysis and liquid–liquid extraction. Thus, important steps in sample preparation can be integrated into the FIA technique. In addition to photometers, electrochemical detectors in particular and also direct potentiometry with ion-selective electrodes are used as detectors (see J, K and L).

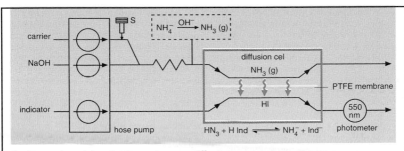

G. Ammonium determination using gas diffusion

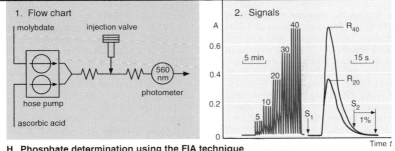

H. Phosphate determination using the FIA technique

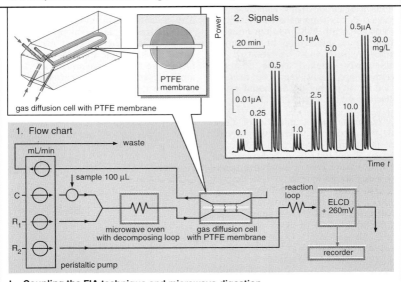

I. Coupling the FIA technique and microwave digestion

J. FIA system with fluoride electrode.
1. Flow chart. With high sample volumes (see D), the dispersion = 1; there is no mixing with a reagent stream. In this case, the FIA system functions as a transport system, namely to a detector, e.g. to an AAS atomizer (see p. 94) or to an ion-selective electrode, as shown here. This characterizes another function of the FIA technique in the framework of modern automated analysis. The fluoride electrode (with an LaF_3 monocrystal) described on p. 58, C, can be used to perform analyses in the range 50 to 200 μg/L. The carrier stream C contains a standard solution with 100 μ/L fluoride. In this manner the response time of the electrode for the fluoride measurement in samples which are injected into this carrier stream is reduced. The reagent streams R_1 and R_2 contain an electrolyte buffer (see Fluoride pH window, p. 59, C.2) or an electrolyte.

On the one hand, the electrode module (ion-selective electrode or ISE chamber) serves as a chemifold for mixing the carrier/sample stream with the electrolyte buffer, and on the other hand, as an electrode holder for the ion-selective electrode and a reference electrode. Both electrodes are arranged in the form of a wall-jet flow-through cell (see p. 158), in which the entire sample flows directly on to the sensor surface. A 3 mol/L KCl solution (R_2) is continuously supplied to the reference electrode. A constant liquid level is maintained by suctioning off the liquid which collects in the ISE chamber.

2. Calibration curve. Because of these special conditions, negative signals (potentials) are observed for fluoride levels between 50 and 100 μ/L. A linear section of the calibration curve is only seen in the region up to about 600 μg/L. The FIA devices are connected to a microprocessor-driven integrated evaluation component which makes it possible to evaluate signals for the height, area or signal width. Both linear (linear regression) and non-linear (Lagrangian interpolation) standard curves can be used for the calibration.

K. Liquid–liquid extraction with an FIA system. Often liquid–liquid extractions are used to separate slightly polar organic substances (see p. 140). These extractions can be performed continuously using the FIA technique thus: the sample is injected into an aqueous carrier stream, and then reagents can be mixed in (R_1, R_2); they can be for the formation of extractable compounds or to mask interfering substances. Then the aqueous stream is segmented by an organic solvent (R_{org}), and the phase transition takes place in an extraction loop after the segmentor. Because of the instability of pump tubes, the organic solvent can also be transported using the displacement principle, as shown here. For example, chloroform can be displaced by pumping water from a vessel into the FIA system. PTFE membranes are used for the phase separation in the separator. A PTFE strip is led from the inlet to one of the outlets of a T-shaped piece of glass. The T-piece is arranged so that the organic phase (with greater affinity to hydrophobic PTFE) is directed downward.

L. Dialysis cell and reducer for nitrate determination in sewage water. Here the dialysis cell has the function of on-line sample cleaning. Different pH values are present in the carrier streams C_1 and C_2; the nitrate ions migrate through the dialysis membrane into carrier stream C_2, larger molecules remain in carrier stream C_1, and therefore they do not reach the reaction system where the reagents R_1 (sulfamide) and R_2 (N-naphthylethylenediamine) are mixed in. Previous to this, nitrate is reduced in the reducing column to form nitrite, which is then converted into an azo dye.

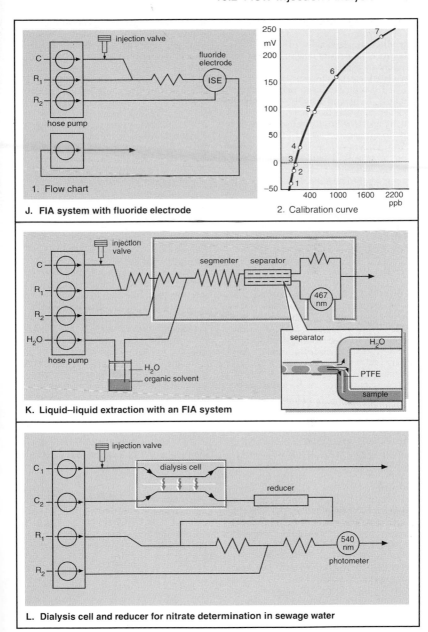

1. Flow chart

J. FIA system with fluoride electrode

2. Calibration curve

K. Liquid–liquid extraction with an FIA system

L. Dialysis cell and reducer for nitrate determination in sewage water

10.3 Coupling Techniques

Introduction. Especially efficient analytical procedures result when methods with different measuring principles and therefore also different selectivity are linked or coupled instrumentally to one another. A coupling technique usually involves an interface between the analytical devices, which can otherwise also be used separately to perform analyses.

The linking of a UV/VIS spectrophotometer with a liquid chromatograph does not represent a coupling technique, because on the one hand, liquid chromatography without detection does not represent a complete analytical procedure, and on the other hand, no special interface is required to link the two devices. In contrast, linking liquid chromatography and mass spectrometry does require an interface, so the LC–MS combination represents a coupling procedure.

A. LC–TLC coupling. By coupling two liquid chromatographic separation methods, it is possible to perform 'two-dimensional' LC: reversed-phase chromatography plays the greatest role in column chromatography and adsorption chromatography does the same in thin-layer chromatography. Using coupling, both separation principles can be linked to one another in one analytical process. The eluate from the HPLC column is applied to a TLC plate via a capillary. The interface in this case consists of a gas supply (e.g. heated nitrogen) and a mechanical system which transports the TLC plate further along as programmed.

B. LC–GC coupling. Even more different are the separation principles which are linked with one another in LC–GC coupling, likewise in the form of 'two-dimensional' chromatography. LC takes the function of automatic sample preparation: the pre-separation, the isolation of analyte components from a complex matrix or trace enrichment. GC is used as a high-resolution separation method (with capillary columns). The interface between LC and GC has the function of transferring an exactly defined fraction from the eluent flow of LC into GC (loop technique). The subsequent retention gap technology uses an uncoated precolumn (retention gap) which is flooded with the contents of a sample loop from HPLC. By switching the valve, the loop is filled with the fraction which is to be analyzed by GC. Then the eluent flow is redirected again and, using the carrier gas stream, the contents of the loop are passed into the GC column. After the sample has been applied, the carrier gas stream is diverted directly to the T-shaped section in order to remove final residues of the eluent in the sample loop backwards through a small opening in the tubing between the control valves.

C. GS–AES coupling. By coupling a capillary gas chromatograph with an atomic emission detector, element-specific chromatograms can be obtained, e.g. for substances containing Cl, N, P and S. The simultaneous measurement of several elements and the recording of emission spectra are also possible. The substances separated by GC are brought via a heated transfer line (a capillary tube) into the atomic emission detector, where the dissociation into atoms and their excitation take place in a microwave-induced plasma (see p. 96). The emitted light falls on a photodiode array via a holographic grid. In this way one to six element lines can be measured concurrently and spectra can be recorded (see also p. 229, F.3).

D. LC–MS coupling. With the particle-beam interface (1), an aerosol is created with a nozzle in an aerosolization chamber, and vacuum pumps are used to remove the eluent from the chamber before the sample molecules enter the ion source. The thermospray process (2) creates gas-like particles, molecules and ions, from which the molecules are suctioned off before they enter a quadrupole mass spectrometer.

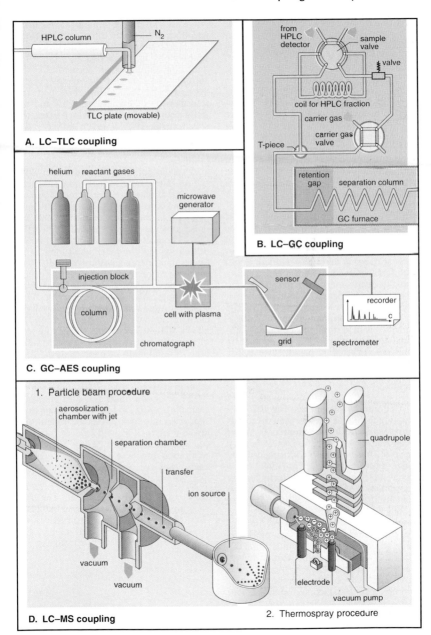

A. LC–TLC coupling

B. LC–GC coupling

C. GC–AES coupling

D. LC–MS coupling

E. LC and AAS coupling. The coupling of liquid chromatography and atomic spectrometry represents a connection between organic and trace element analytical procedures. With the coupling technique shown, for example, protein fractions can be determined in which certain metals (e.g. Se, As and Bi) are bound. A gel chromatographic separation using a low-pressure liquid chromatograph is followed by UV detection to detect the proteins at 205 or 254 nm. The coupling technique itself consists of a microwave digestion system and an FIA module. In this interface the proteins are partially degraded after the addition of acid (supplied via the FIA module). After the addition of sodium borohydride, the conversion of the metal compounds (complexes) into hydrides takes place in the FIA module. The function of the microwave degradation in series is to prevent foaming during the reduction and therefore to prevent proteins from entering into the quartz measuring cuvette of the atomic spectrometer. Subsequent AAS analysis occurs after atomization in the quartz cuvette with flameless methodology. Signals obtained using this coupling technique can be displayed over one another by computer, making it possible to assign selenium, bismuth or arsenic levels to certain protein fractions in an automated procedure.

F. ICP and MS coupling. The ICP–MS coupling technique uses an inductive coupled plasma as an ionizing source, as is used in optical atomic emission spectrometry (see p. 92) in conjunction with a quadrupole mass spectrometer (see p. 124). Mass spectrometry is used in this form for the determination of elements, whereby a high absolute detection level of down to 1 pg can be reached. An atomizer (1) is used to transport (2) the sample with argon into the analytical zone of the plasma produced using a high-frequency generator. From the inductively coupled plasma burner, out of the analytical zone, the ions formed there are transported to a gap with the help of a sampler, a cone-shaped perforated screen. Using a second perforated screen, the skimmer, a portion of the ions are separated

out in this gap and are brought into the mass spectrometer, in which the working pressure is less than 10^{-5} mbar. This procedure is also called ion extraction.

Several ion lenses are located after the skimmer. By changing the voltage current applied there, one can optimize both the transmission of the spectrometer and the mass resolution for an ion which enters the spectrometer with a given energy. The vacuum in the mass spectrometer is produced using a diffusion pump or a cryopump (3).

A quadrupole mass spectrometer consists of four parallel rods at equal distances from one another, to which a direct current field and also a high-frequency field are applied (4). Electron multipliers and the pulse counting process are used to detect the ions. The ICP mass spectrometer is controlled by the computer connected to it. Spectral interference occurs particularly in the mass range up to 80, as a result of ion formation from water and from the acids of the sample solutions and of argon. This makes it more difficult to determine light elements in some cases. However, on the whole, ICP–MS combines the advantages of a simple sample input with the possibilities of rapid multi-element analysis, with high detection sensitivities for almost all elements and the ability to use isotope dilution analysis (see p. 200).

The detection limits for most elements are in the range 0.1 to 1 ng/mL—on average approximately one order of magnitude lower than in ICP with optical (atomic) emission spectrometry (see p. 96). Linear calibration functions can be achieved in many cases over 4–5 powers of ten.

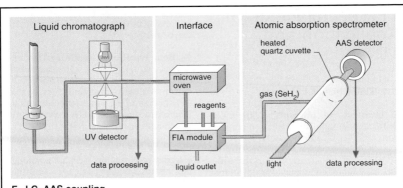

E. LC–AAS coupling

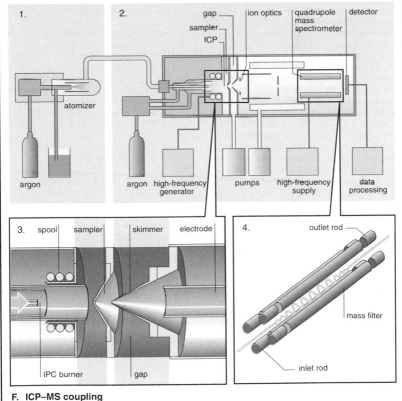

F. ICP–MS coupling

11 Special Fields of Application and Methods

11.1 Radiochemical Methods

A. Overview of methods. With radiochemical methods, the radiation emitted from radioactive atomic nuclei is used as analytical information. The activation of atoms occurs principally due to the influence of high-energy particles or quanta, whereby nuclei with a large energy excess are generated. These nuclei are therefore metastable and return back to low-energy nuclei by giving off particles or quanta. If the radioactive nuclei are generated in the analytical sample itself (using a neutron source: reactor or radionuclides), we are dealing with an activation analysis. The energy and intensity of the disintegration radiation are measured so that this neutron activation analysis makes possible both qualitative and quantitative multi-element analyses (A = element mixture) even in complex matrices (M; see also C). Instrumental neutron activation analysis (INAA) uses energy-dispersive spectrometers; the samples can be analyzed without interference. However, if counter tube detectors are used, then chemical separations must occur prior to the analysis. Radiochemical analysis or tracer analysis uses radioactive nuclides as highly sensitive measurable indicators. In isotope dilution analysis, a small, exactly defined amount of the radioactive material is added to the analytical sample. In trace element analysis, stable isotopes are used, and the isotope distribution is determined subsequently using high-resolution mass spectrometry. In direct isotope dilution, one adds a radioactive isotope of the element to be determined to a weighed amount of sample, then one separates an arbitrary amount of the desired substance, along with its added isotope in pure form, perhaps by precipitation or extraction, and one determines the activity of this portion. In this way, not only can very small amounts be detected, but also process steps (recoveries) in trace analysis can be monitored.

B. Beta scintillation counter. To detect radioactive radiation, in this case β-radiation, scintillation counters are used as detectors in radioimmunoassays (see also p. 48). The sample, which contains tritium-labeled substances, emits β-rays (electrons) which land on the scintillator (inorganic or organic luminous substances) and produce flashes of light which are converted into electron impulses on the photocathode of the photosecondary electron multiplier. The system of parallel electrodes (dynodes) produces an increase in the current impulse by a factor of 10^6 to 10^8 based on the principle of secondary emission.

C. Neutron activation analysis. Neutron activation analysis represents a particularly sensitive method, with nuclear reactors as the most important neutron source. The amount of radionuclide produced is proportional to the neutron flow, the number of nuclei capable of reacting and the capture cross-section (as a measure of the probability that a nuclear reaction will start). In the best cases, detection limits in the lower picogram range can be achieved. The example shows the result of a sea water analysis in which the neutron activation analysis makes possible the qualitative and quantitative analysis of iron, cobalt, zinc, sodium and chlorine (as chloride).

D. Radio-thin-layer chromatography. The iron isotope ^{59}Fe was added to coffee and tea extracts (in ethyl acetate) in the context of element speciation analysis (see p. 220). After separating the organic substances on silica gel, the iron-binding fractions can be selectively detected by radiography. A methane flow counter (γ-rays = X-ray quanta of ionizing carrier gas) is used for the detection. A noticeably broader distribution of the iron is found in the coffee extract compared with the tea extract.

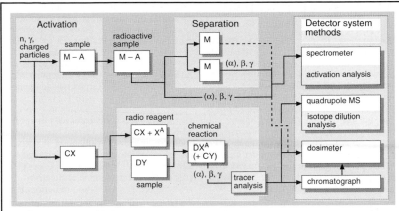

Activation

n, γ, charged particles → sample M – A → radioactive sample M – A

Separation

M

M   (α), β, γ

(α), β, γ

radio reagent

CX + X^A

DY

sample

chemical reaction

DX^A (+ CY)

(α), β, γ → tracer analysis

Detector system methods

spectrometer

activation analysis

quadrupole MS

isotope dilution analysis

dosimeter

chromatograph

A. Overview of methods

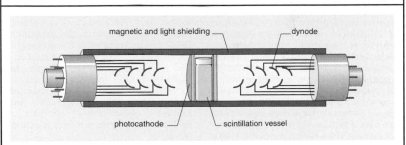

magnetic and light shielding — dynode

photocathode — scintillation vessel

B. Beta scintillation counter

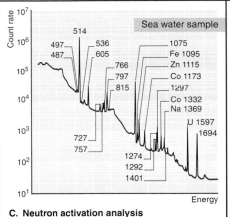

Count rate

10^7
10^6
10^5
10^4
10^3
10^2
10^1

Sea water sample

514
497   536
487   605
766
797
815
727
757
1075
Fe 1095
Zn 1115
Co 1173
1297
Co 1332
Na 1369
U 1597
1694
1274
1292
1401

Energy

C. Neutron activation analysis

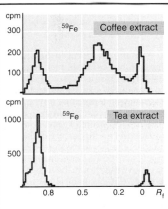

cpm
300
200
100

^{59}Fe   Coffee extract

cpm
1000
500

^{59}Fe   Tea extract

0.8   0.5   0.2   0 R_f

D. Radio-thin-layer chromatography

11.2 Solid Body and Surface Analysis

Introduction. In the analysis of solids, mostly non-destructive methods are used, the results of which are used to answer questions as to the elemental composition in microregions, the bonding conditions of the elements detected and the change in composition as the depth increases, for example. Methods for surface analysis analyze a small layer of the solid body (information depth) but which has finite thickness. By stepwise or continuous ion bombardment, called sputtering, each analyzed layer can be stripped away and thus a depth analysis can be performed with high resolution.

A. Principles of secondary ion mass spectrometry (SIMS). The SIMS procedure links the bombardment of a solid surface by ions, the primary ions (with energies up to 20 keV), with mass spectrometric analysis of the ionized particles (the secondary ions) which split off in the process. The primary ion beam of positively charged argon, cesium, oxygen or nitrogen ions (or also negative oxygen ions) is focused and directed at a selected area of the solid sample (1). Based on their mass number, the particles which are stripped ('sputtered') are captured mass spectrometrically (see p. 126) using a double-focusing (DF) mass spectrometer (2). The depth distribution (3) of elements can be obtained by recording the appropriate secondary ions during the sputter process.

B. Signal generation in SIMS. Gas-discharge tubes or liquid metal systems (for cesium) are used as ion sources in which the ions produced can be removed from the plasma using a potential on the extraction electrode and can be directed into a lens system. The primary ions from the ion source meet on the surface of the solid and transfer impulses to the atoms there (1). The target (solid body) atoms change location as they move lower or in the form of sputtered particles (largely neutral particles, with a few ions). The depth-resolving capacity amounts to approximately 3 to 5 atomic positions. Mono- or multivalent atoms are measured for elemental analysis and molecule or cluster ions are measured for distribution-specific analysis. A section from the mass spectrum (2) of a SIMS analysis of a hard metal using positive oxygen molecule ions as primary ions with an energy of 5.5 keV shows the presence of various positively charged ions in the mass region up to 220.

C. Auger electron microanalysis spectra (AES). Auger electron spectrometry includes the measurement of the kinetic energy of electrons which are emitted by a substance as the result of the Auger effect (see p. 98). Electromagnetic, electron or ion rays are used as activation radiation. If the surface of the sample is bombarded by electrons, secondary electrons are released. Owing to the electric or magnetic fields in the region between the sample and the collector, only electrons with a particular energy reach the aperture. By changing the fields, electrons having different energy are focused on the detector, one after another. The measured intensity is recorded as a factor of the field strength (1). As a rule, an incandescent cathode is used as the excitation source. Element-specific Auger electron spectroscopy provides an informational depth of only a few atomic positions.

The secondary electron spectrum (2) of the surface of a German silver alloy (made of Cu–Ni–Zn) annealed at 730 °C for 10 min, recorded with an Auger probe, shows in particular the assignment of the Auger electrons to the elements sulfur, chlorine, oxygen and carbon. On the other hand, the main components copper and nickel cannot be detected on the surface. Thus, the surface composition differs decisively from the composition of the alloy.

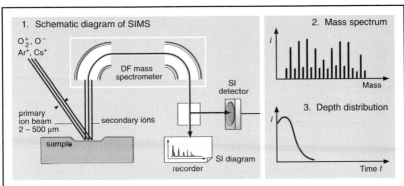

1. Schematic diagram of SIMS

O_2^+, O^-
Ar^+, Cs^+

DF mass spectrometer

SI detector

primary ion beam 2 – 500 μm

secondary ions

sample

SI diagram

recorder

2. Mass spectrum

I

Mass

3. Depth distribution

I

Time t

A. Principles of secondary ion mass spectrometry (SIMS)

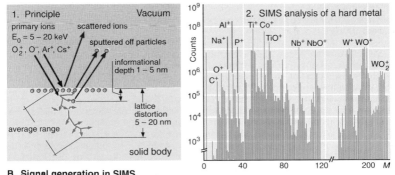

1. Principle   Vacuum

primary ions
$E_0 = 5 – 20$ keV
O_2^+, O^-, Ar^+, Cs^+

scattered ions

sputtered off particles

informational depth 1 – 5 nm

lattice distortion 5 – 20 nm

average range

solid body

2. SIMS analysis of a hard metal

Counts

10^9
10^8
10^6
10^5
10^4
10^3

Al^+
Na^+
O^+
C^+
P^+
$Ti^+ Co^+$
TiO^+
$Nb^+ NbO^+$
$W^+ WO^+$
WO_2^+

0   40   80   120   200 M

B. Signal generation in SIMS

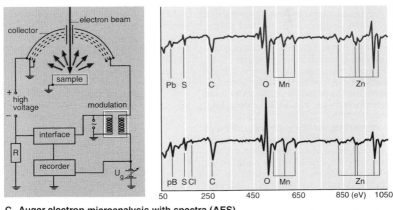

electron beam

collector

sample

high voltage

interface

modulation

R

recorder

U_g

Pb S   C   O   Mn   Zn

pB S Cl C   O   Mn   Zn

50   250   450   650   850 (eV) 1050

C. Auger electron microanalysis with spectra (AES)

D. Principles of photoelectron spectrometry. Owing to the effect of X-rays or UV radiation on solid bodies, electrons are emitted as a result of the photoelectric effect, and the kinetic energy of these electrons is measured. The term 'electron spectroscopy for chemical analysis' (ESCA) was coined for this method. In addition to X-rays, UV radiation from gas-discharge tubes can be used as an excitation source. Photoelectron spectrometry is suitable for the nondestructive analysis of solid surfaces and of the liquid or gaseous substances which adsorb to them. Being dependent on the primary energy, photoelectron signals are element-specific. If zirconyl dimethacrylate radiation (with 151 eV) is used, it is also surface specific, since the emission depth of the photoelectrons reaches a minimum at a kinetic energy of about 100 eV.

E. Structure of an X-ray photoelectron spectrometer. An X-ray photoelectron spectrometer consists of an X-ray source, the sample chamber, the energy-dispersive system for generating monochromatic X-radiation and the analyzer and detector systems. If magnesium or aluminum are used as anode materials of the X-radiation source, the K_α rays have half-widths of about 0.7 to 0.8 eV, and often measurements can be made without a monochromator. A monochromator reduces the linewidth down to a few tenths of an eV. Different analyzers can be used to measure the kinetic energy of the emitted photoelectrons, such as an electrostatic counterfield or an analyzer shaped like a semicircle with a magnetic and an electrostatic field. The signals from the analyzer which reach the detector are counted or, with continuous variation of the fields, they can be converted into a counting rate by an integration unit.

F. Photoelectron spectra of different copper surfaces. The spectra in the figure show the surface of a sample of pure copper, of a sample contaminated by exposure to the air and an additional oxidized sample, all with the same device parameters. Surface coating with hydrocarbons (pump oils), oxides and adsorbed water vapor led to a weakening of the actual spectrum. For this method, an informational depth of 1 to 10 nm is indicated.

G. Laser spectrometry. *1. Overview.* Lasers produce monochromatic light rays which are very intensive and sharply bundled (laser = light amplification by stimulated emission of radiation). These rays can be used for sample vaporization and for bulk and micro-region analysis. The elements which are present in the form of atoms or ions after the sample has been vaporized with a focused laser beam can be transported for detection using atomic spectrometric methods such as atomic absorption spectrometry (AAS), optical atomic emission spectrometry (OES), laser spectrometry itself (LEI: laser enhanced ionization) or mass spectrometry.

2. Principles. The atomic cloud generated by sample vaporization by laser is irradiated by a laser, whereby the laser's frequency is synchronized with the resonance line of the element to be analyzed (analyte). The three basic principles are: absorption of the radiation by the analyte (LAAS: laser atomic absorption spectrometry), fluorescence (laser-induced fluorescence: LIF) or the production of ionization products (ions and electrons). Regarding this ionization, one differentiates between thermal impact with other particles (LEI: laser-enhanced ionization), ionization due to an electric field (FILS: field ionization laser spectrometry) and that due to photoionization (RIS: resonance ionization spectrometry). This type of ionization can be combined with a mass spectrometer to detect isotopes selectively (RIMS: resonance ionization mass spectrometry).

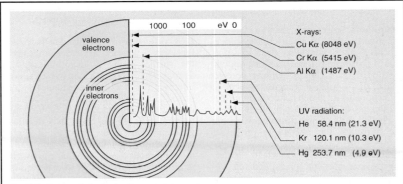

D. Principles of photoelectron spectrometry

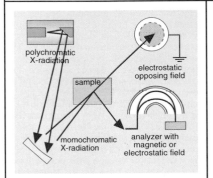

E. Structure of an X-ray photoelectron spectrometer

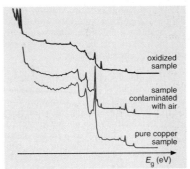

F. Photoelectron spectra of different copper surfaces

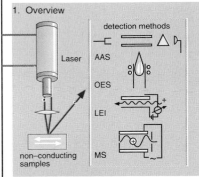

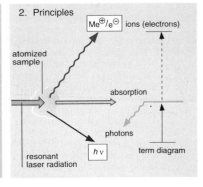

G. Laser spectrometry

11.3 Chemical Sensors

Definitions. Sensors are detecting elements which receive optical, electric or substance-specific signals and convey them in a processed (converted) form (see also special electrodes on p. 56). Chemical sensors consist of a substance-recognizing portion and a transducer, which converts the information into an electric or optical signal.

Physical and chemical mechanisms are the basis for the interactive processes between the analyte and sensor. These mechanisms will be described in the following examples for special sensors frequently used in actual practice: physisorption with relatively weak compounds, e.g. van der Waals forces (see also p. 134) and chemisorption, volume intercalation and surface, interface and three-phase reactions. Of these, interface reactions play a key role in the use of special materials for chemical sensors.

A. Principles of a chemisorption sensor.
Electric and optical properties of a sensor are often changed due to (selective) chemisorption (1). For example, the stationary surface conductivity on an SnO_2 chemisorption sensor changes as a factor of the NO_2 partial pressure in the air. The conductivity of the material is affected by the adsorption of NO_2 molecules and of the compounds resulting from this action. Basically, the conductivity increases or decreases, depending on whether free electrons are produced or removed. When an NO_2 molecule binds, the analyte takes up a conduction electron from the SnO_2 surface and the stationary surface conductivity is reduced. The reduction of the surface conductivity or an increase in the electron affinity on the surface (2) can be explained by the generation of a surface dipole, in addition to the electron capture which we mentioned. Cluster calculations or band diagrams (1) are used to describe the electron transfer to n-type semiconductors. Semiconductor sensors are generally used for measuring the concentration of oxidizing or reducing gases. The conductivity changes as a result of the accumulation of the gas molecules on the surface of homogeneous semiconductor gas sensors.

B. Oxide sensors. With interface and three-phase sensors, one uses changes near the surface using a location-dependent dosing. The Schottky diode has a Pt/TiO_2 interface (1). TiO_2 sensor structures contacted this way demonstrate a metallic binding of Pt atoms to their surface at lower temperatures. In this form, they display a gas-dependent diode behavior. If higher temperatures and higher oxygen partial pressure both occur simultaneously, the relatively large metallic Pt atom is oxidized on the surface, then the smaller Pt ion is incorporated between the first and second atomic positions of TiO_2 surfaces. The electron acceptor, platinum, becomes an electron donor, Pt ion. After the diffusion of Pt ions into TiO_2, the diode characteristic curve is converted into an ohmic curve. The sensor characteristic corresponds to Ohm's law with respect to electric resistance. The oxygen ionic conductivity is also used with the lambda probe (see C) as a sensor principle. At the Pt/ZrO_2 phase transition there is a conversion of O_2 (gas phase) into O^{2-} ions (in ZrO_2), and the Nernst potential is measured. The fundamental sensor principles of physisorption, chemisorption, surface defects or volume defects with oxide sensors are dependent on the temperature (2) or are preferentially effective in particular temperature ranges.

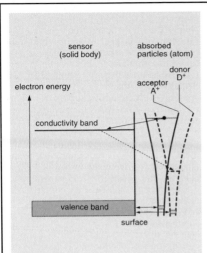

1. Band model and energy spacing diagram

A. Principles of a chemisorption sensor

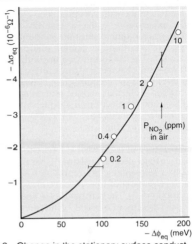

2. Change in the stationary surface conductivity $\Delta\sigma_{eq}$ and in the electron affinity work function $\Delta\phi_{eq}$ as a function of the NO_2 partial pressure on SnO_2 chemisorption sensors at 450 K

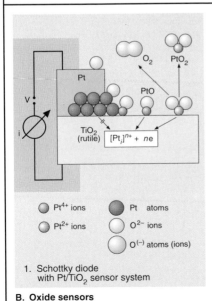

1. Schottky diode with Pt/TiO$_2$ sensor system

B. Oxide sensors

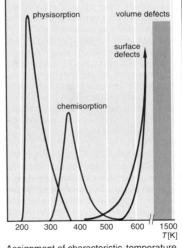

2. Assignment of characteristic temperature ranges to the particular sensor functions with oxide sensors

C. Potentiometric oxygen sensor. With the lambda probe with yttrium-doped ZrO_2 as the oxygen ion conductor (see B), the analyte gas mixture reaches the solid electrolyte surface through porous platinum electrodes. The cathode reaction (both in the measuring chamber and in the reference chamber) leads to an equilibrium between the oxygen molecules and the grid oxygen (as O^{2-}) and the electrons in the platinum. There is a constant oxygen partial pressure in the reference cell.

D. Structure of a CHEMFET (chemical sensitive field effect transistor). Chemical sensitive semiconductors represent a further development of ion-selective electrodes (Section 5.3). They are made of n- and p-silicon, whereby as a result of the manufacturing process, two regions enriched with n-Si arise in the p-Si (largely covered by SiO_2). Si_3N_4 serves as a stable protective layer, the actual ion-selective layer is between the n-Si zones, and aluminum electrodes are coated at the points 'source' and 'drain,' which can be used to influence the two n-Si regions galvanically. A current flow in the gate area (trigger electrode of the FET) occurs only at a particular drain potential U_D. This current flow is affected capacitatively by an additional electric field in the gate area. Finally, the gate potential U_G, which controls the drain stream, is composed of the potential U_G (applied to the reference electrode) and the potential jumps at the phase boundary between the ion-selective layer and the electrolyte. The drain stream is affected by the dipole formation at the boundary layer between the ion-selective layer and the electrolyte.

E. Optical sensors. The development of optical sensors is based on glass-fiber optics in conjunction with chemical systems. The *in vivo* optosensor based on fiber optics (1) consists of a small cellulose dialysis tube in which the microencapsulated indicator phenol red (bound to polyacrylamide particles) and polystyrene beads approximately 1 μm in size—as a filling—are found. The dialysis tube is directly coupled to a pair of light guides made of plastic. The irradiated light is scattered on the poly-

styrene beads; the indicator beads, which are in contact with body fluids such as blood, change their light absorption behavior as a function of the pH. Finally, the light reflected via the second light fiber is transferred to a photodiode. Fiber-optic sensors can also be used in flow systems (FIA, see Section 10.2). A suitable pH optosensor (2) contains the indicator immobilized on porous material. The changes in absorption are measured using remission analysis (see Section 9.2.L). The ammonia optosensor (3) makes possible continuous measurement as a result of the separation of the analyte stream from the indicator system using a white, permeable Teflon membrane. The (stationary) color change which occurs at the interface (as a function of the current NH_3 concentration) is continuously measured using fiber optics. Light guides can also be directly coated with a chemical coating, one end of which is irradiated by the light from a luminous diode and the other end of which is a recording photocell. The irradiation of the gas-sensitive coating is done while maintaining the critical angle for an internal reflection at the polymer/air boundary layer. If analyte molecules diffuse into the layer and react with the reagent present there, there is a weakening of the internally reflecting light beam.

C. Potentiometric oxygen sensor

heater

conducting electrodes (porous Pt)

solid electrolyte (Y_2O_3/ZrO_2)

measuring chamber

reference chamber

C_i C_a

anode reaction
$$2O^{2-} \longrightarrow O_2 + 4e^-$$

cathode reaction
$$O_2 + 4e^- \longrightarrow 2O^{2-}$$

$C_a > C_i$

U

thermo-element

anode — + cathode

D. Structure of a CHEMFET

U_g

reference electrode

isolation

drain

electrolyte

IS layer

SiO_2 gate

n-Si   n-Si   p-Si

SiO_2
Si_3N_4
source (Al)

n-channel

U_D i_D

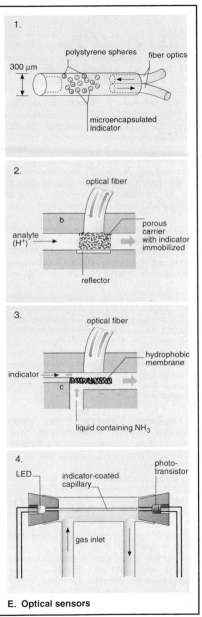

1.

300 µm

polystyrene spheres   fiber optics

microencapsulated Indicator

2.

optical fiber

b

analyte (H^+)

porous carrier with indicator immobilized

reflector

3.

optical fiber

indicator

hydrophobic membrane

c

liquid containing NH_3

4.

LED

indicator-coated capillary

photo-transistor

gas inlet

E. Optical sensors

11.4 Process Analysis

Definition. Process analysis is used in the monitoring and control of chemical processes. The measurements are made in a time-dependent manner, so these are dynamic analyses.

A. Signal forms. *1. Continuous and discontinuous signals.* Analytical equipment to be used in process analysis should determine the concentration, e.g. of individual components as a function of time. One differentiates between continuous (a) and discontinuous (b) signal forms. Depending on the type of y–t recorder, analog signals can be recorded continuously or discontinuously (with a point recorder).

2. Concentration–time functions and scanning frequency. Depending on the individual values measured, high (A, B with strong and rapid changes) or low scanning frequencies (C) are necessary.

3. Dead time and time constants. Furthermore, depending on the analytical equipment, one must take into account a time shift between the current and the measured concentration—as a result of the material transport between the sampling and the measuring sites. A discontinuous change in the measured value at time t_0 is indicated by a delay (time constant t defined as 63% of the final measured value).

B. Equipment diagram for paramagnetic oxygen measurement. The property of paramagnetism is used for process analysis in gases as if it were specific for oxygen. The gas analyte is led through a ring-shaped chamber. Both halves are connected by a tube with two filament windings as part of a Wheatstone bridge switch. The magnetic susceptibility (capacity) is larger in the left (cold) part of the tube near the asymmetrically positioned permanent magnet than in the right (hot) part, so that the oxygen is drawn in more strongly by the left part. The flowing gas transports heat, whereby the resistance of the two filament windings changes.

C. Process pH meter. As process devices, operative pH meters have a one-rod glass electrode, a resistance thermometer and a temperature compensation device. An antimony electrode (see p. 56) is more stable than a glass electrode, both mechanically and electrically, but it cannot be used in strongly oxidizing or reducing media.

D. Process refractometer. Light from a sodium vapor discharge lamp lights up a gap monochromatically. From there the light goes to two adjacent, hollow prisms, one of which is filled with a reference fluid. The left prism represents the flow-through cell. The difference in the refractive indices is determined. The function of the mirror is to compensate for the deflection of the light beam so that it is not the photoelectric current itself which is used, which is kept constant, but rather the position of the mirror which is used for the determination of the concentration of a substance (e.g. in brewing beer or in the fruit juice industry, monitoring sugar levels, etc.).

E. Process photometer. By using a double-beam photometer, one can compensate for unclean windows and aging of the light source with the differential measurement procedure. To do so, measuring light from a deuterium lamp might be split alternately into the measuring and reference path by a mirror oscillating at a frequency of 500 Hz (principle of an optical bridge). The light reducer is used to change the intensity in the reference beam until both intensities are equal. The position of the light reducer of the display is a measure of the transmission. Dirty windows are compensated for owing to the differing layer lengths in the measuring and reference beams (P1, P2). The difference between the two cuvette layers yields the effective layer thicknesses.

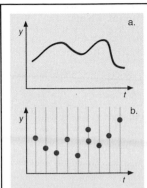

1. Continuous and discontinuous signals

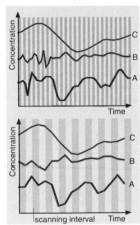

2. Concentration–time functions and scanning frequency

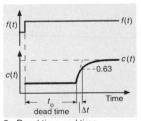

3. Dead time and time constants Δt

A. Signal forms

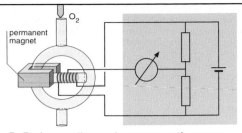

B. Equipment diagram for paramagnetic oxygen measurement

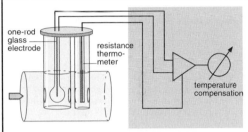

C. Process pH meter

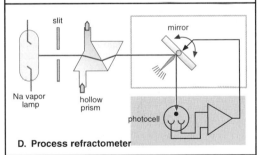

D. Process refractometer

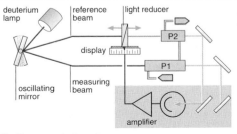

E. Process photometer

F. Procedural steps in process analysis. Starting with the process to be monitored, the steps involved in process analysis include sampling, sample preparation (often with a high degree of difficulty), the actual analysis or measurement and signal detection, signal conversion and transfer to a processor, which can also direct the process.

Usually in-line processes can be used for the continuous detection of physical values such as pressure, temperature, conductivity, refractive index or flow throughput. 'In-line' means that the sensor can be placed directly in the sample stream. For industrial process monitoring, usually on-line analyzers with integrated sampling and sample preparation are used to determine chemical parameters. Sample preparation includes steps such as sample filtration, controlling the sample temperature, dilution and digestion. In sewage water process analysis (see G), the commercially available devices which have been developed for this purpose require low-sulfur samples in order to avoid deposits and clogging in the narrow tubes and valves, and also to avoid interference during the analysis. Continuously working filter and screening systems are used for this, which can be cleaned with water under pressure or by using ultrasound. Even the analysis of gases usually requires the separation of the dust particles before the actual measurement. Mostly spectrometric methods are used for the analysis itself in both gas and water analysis.

An important aspect of process analysis is the processing of the information. Here is a method for achieving high reliability in information transfer and in the execution of operational functions in conventional systems: all measuring and analysis devices are arranged in parallel; they transmit and process information in analog form in a central control room. However, today, as a rule computer processors are used, which can use programs to produce time-related connections from all of the data which are recorded or manually input. In this way, at the highest level of process automation, one can even directly exert control over the process.

G. Continuous UV measurement in sewage treatment plants. The overall detection of the levels of organic material in effluent in the form of the biochemical oxygen demand (BOD) (see p. 216) is one of the basic quantities to be measured for the optimal operation of a sewage treatment plant. However, BOD determination takes considerable time and is not well suited for automation. In many cases there is a good correlation between the UV absorption at a particular absorption wavelength. This analytical method can be drawn upon for the continuous monitoring of levels of organic, effluent-burdening substances.

The flow chart (1) depicts such an analytical system: a peristaltic hose pump pumps distilled water as a blank solution through a UV flow photometer (see E). A three-way solenoid valve which can be controlled by a controller unit is switched for brief periods to the sample stream from which suspended matter has been removed by a membrane filter. In this way, a defined sample volume enters the flow system (see FIA, p. 190) and into the detector. The control unit switches the three-way valve via an adjustable signal generator with a relay output, and at the same time it activates the auto-zero function of the UV detector. The input signals (2) leaving from the detector and selected by the control unit are captured in analog form, are stored and are recalled by the data processor at regular intervals. After the sample has been filtered, systems for the determination of phosphate, ammonia or nitrate can be connected.

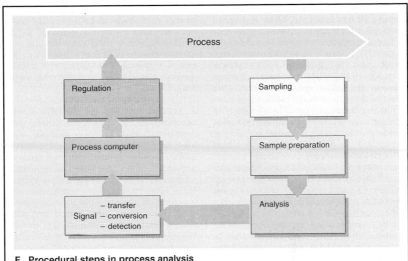

F. Procedural steps in process analysis

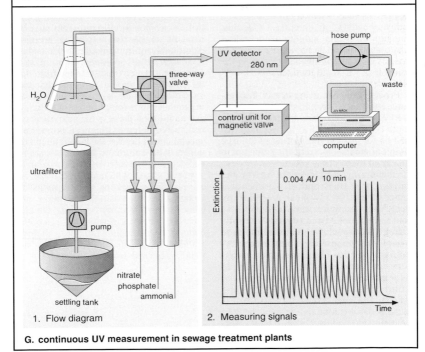

1. Flow diagram

2. Measuring signals

G. continuous UV measurement in sewage treatment plants

H. Functional principles of a non-dispersive IR gas analyzer. There is a broad application for the absorption measurement of IR radiation (see Section 8.5) in the range from 20 to 1 μm or 500 to 10000 cm^{-1} for gases such as aliphatic hydrocarbons, ammonia, water, carbon monoxide, carbon dioxide and sulfur dioxide. There is no spectral dispersion of the radiation in non-dispersive IR gas analyzers. In a device with positive filtering, the polychromatic radiation of two rays with the same intensity is modulated and alternatively directed through the measuring cuvette and a reference cuvette filled with IR-inactive nitrogen. Two chambers filled with the analyte gas (e.g. CO) and separated by a membrane condenser serve as radiation receivers. If a difference in radiation absorption occurs in the two cuvettes, fluctuations occur in the temperature and therefore in the pressure in both receiving chambers in rhythm with the modulation. The condenser membrane is deflected to differing degrees, resulting in changes in capacity which can be measured and, after calibration, output as CO levels. The high long-term stability of the analytical system which is required in process analysis is made possible here by using the principle of varying light intensity.

I. Process analysis using X-ray fluorescence analysis. X-ray fluorescence analysis (XRFA; see p. 100) plays an important role in the cement industry as a discontinuous method for process monitoring with large sample numbers—as a detecting element in the closed loops of the production process for monitoring sulfate levels (i.e. of the addition of plaster) and of the chemical composition of the raw meal. After the raw meal mill, samples are automatically drawn, then packed and transported to the XRFA spectrometer by a pneumatic tube system after being called up by the central processor (see F). The sample preparation consists of molding the fine powder sample of the raw meal in aluminum cups (without further pretreatment, i.e. without adding binder). The value for the lime standard is calculated from the analytical data and they are compared with a predetermined reference value. If a tolerance limit is exceeded, the process computer calculates the amount of clay that needs to be added and automatically makes the adjustment in the continuous scales for dosing the clay; the amount of limestone is kept constant. This makes it possible to produce a desired quality cement which is stable over a long time.

J. Set-up for process HPLC. Today, automated GC has become a general standard of industrial analysis methods when used as process GC, i.e. for the totally automated performance of serial analyses with bypass valves for direct charging from product stream into the carrier gas system. HPLC can be used for thermolabile and unevaporable materials, e.g. in the field of water studies and biochemical and pharmaceutical manufacturing. As process HPLC, it is technically more involved and must be easy to maintain as well as explosion proof. Even away from locations where there is a risk of explosion, it is necessary to protect against corrosion of the entire control electronics in an aggressive environment as an internal explosion proofing. Additional requirements are: the robustness of all analytical components; optimization for long residence times, as seen from both the column technology and the detection technology; suitable hardware and software in general; control electronics for switching and monitoring tasks in the analytical system; fully automated diagnosis programs; a special eluent device for large eluent containers with automatic bubble separation and degassing; a safety exhaust system for any toxic vapors which might develop; and an evaluation system with options for recognizing decreasing separation efficiency and for introducing countermeasures such as rinsing the separation column. Diagnosis software includes warning messages, safety switches and pump, valve and detector diagnosis.

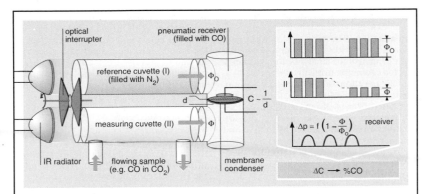

H. Functional principles of a non-dispersive IR gas analyzer

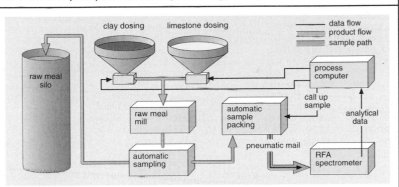

I. Process analysis using X-ray fluorescence analysis

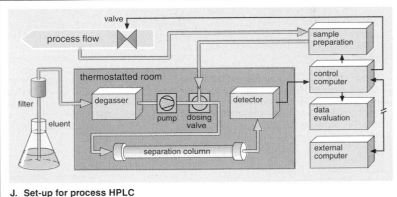

J. Set-up for process HPLC

11.5 Structural Analysis

Introduction. The goal of structural analysis is to provide information about the arrangement of the elements, the symmetry and the geometry, using this information to answer questions about the constitution, configuration, conformation, the complete spatial structure (with interatomic distances and valence states), and electron density distribution for molecules. Qualitative and semiquantitative structural information can primarily be obtained by molecular spectrometric methods.

A. Qualitative and semiquantitative structural analysis. Information about the structural isomerism does not provide any statement about the spatial configuration of molecules. UV/VIS, IR and NMR spectrometric methods include transitions of linkage electrons or systems, oscillations and rotations, as well as magnetic nuclear states preceded by absorption processes. Additional information that is needed includes the empirical formula of a substance ($C_8H_{15}NO_2$ in the example), the molar mass ascertained from the analysis (methods such as raising the boiling point, lowering the freezing point or using mass spectrometry) and elemental analysis.

1. Mass spectrum. In the mass spectrum, the molecular peak indicates the molecular mass, with mass 157. Therefore, with the results from the elemental analysis of 61.1% C, 9.6% H, 8.9% N and 20.4% O, we arrive at the empirical formula $C_8H_{15}NO_2$. The other mass numbers represent molecular fragments, to which different atomic groups can be assigned using tables.

2. UV spectrum. The UV spectrum yields an absorption maximum of 200 nm with a spectral absorption coefficient of 1.259×10^4 ($\log \varepsilon = 4.1$), measured in ethanol. The absorption band points to a structure element with the composition =C=C—COOR.

3. Proton nuclear magnetic resonance spectrum. The information obtained from this method indicates functional groups and

their neighboring groups, so that the assignment of the signals makes clear the nitrogen bonds and a considerable part of the entire molecule, namely the groupings around a carbon–carbon double bond. The two signals which can be assigned to nitrogen–carbon linkages indicate the end terminal ethylene groups and the linkage to two methylene groups as a third group.

4. IR spectrum. Finally, the IR spectrum confirms the carbon–carbon double bond in the form of a CH_2=C bond and the presence of a carbonyl or ester group. With the aid of the IR spectrum, this simple substance can be identified by comparing spectra (using printed or computerized collections of spectra).

The comparison of all information from the different molecular spectrometric methods shows that some molecular structure groups can be derived from several spectra; in this event, one speaks of overlapping information, although it increases the certainty of the assignments at the same time. Therefore, combining different methods is sensible in any case, and for complicated molecules it is a necessity. The example shows work published by Clerc, Pretsch and Seibl [*Chemie für Labor und Betrieb*, **23**, 108 and 158 (1972)] for methacrylic acid β-dimethylaminomethyl ester (also known as colamine methacrylate ester). It shows the pieces of information obtained from the four molecular spectrometric methods and then how they are pieced together to represent the structural formula for this substance, which shows in particular the usefulness of this combination of spectra. The subsets of information complement each other relative to the functional groups, the framework and the symmetry. As in content analysis, performing analyses using methods which are independent of one another owing to their being based on different principles increases the degree of certainty concerning the accuracy of the results.

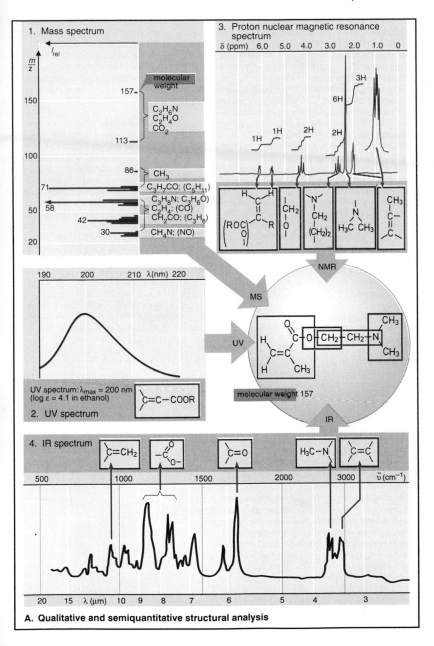

1. Mass spectrum

3. Proton nuclear magnetic resonance spectrum

molecular weight

C_2H_6N
C_2H_4O
CO_2

CH_3
$C_3H_7CO; (C_5H_{11})$
$C_3H_8N; C_3H_6O$
$C_2H_4; (CO)$
$CH_2CO; (C_3H_6)$
$CH_4N; (NO)$

2. UV spectrum

UV spectrum: $\lambda_{max} = 200$ nm
(log ε = 4.1 in ethanol)

molecular weight 157

4. IR spectrum

A. Qualitative and semiquantitative structural analysis

B. X-ray diffraction methods. Complex structures can be clarified using diffraction methods. They provide information about the order, spacing and angles in the structure, especially of crystals. They represent physical–mathematical methods. The instrumental methodology is based on the following principles. The primary radiation directed toward the sample from the radiation source (X-ray tube, neutron or electron source) via diaphragms is elastically dispersed and diffracted at the atoms. The function of the detector, in combination with a monochromator, is to determine the intensity of the diffracted rays as a factor of their spatial position. For example, an X-ray film or a site-sensitive radiation detector can be used to record the intensities. The monochromator is used to shield non-coherent stray components. In the Laue X-ray method, a crystal is radiated with polychromatic X-ray light from almost parallel wave trains. One obtains diffraction patterns which make it possible to make deductions about the ratios of symmetry, the crystal class and the angle ratios between the reflecting lattice plane. In the revolving crystal and goniometer methods, monochromatic X-radiation strikes the crystal. In the second method, the crystal is situated on a rotating goniometer head. Of the radiating X-ray light, the greater portion penetrates the crystal in a straight direction, and a small portion is amplified under a very specific angle, the glancing angle θ of Bragg's equation ($n\lambda = 2d\sin\theta$). This produces a dark line on the film bent in a semicircle. In the revolving crystal procedure, the goniometer is rotated until the glancing angle is reached. Both methods provide information about the type and size of the unit cell (crystallographic class, grid constants) and about the external and internal crystal symmetry (space groups). In the Debye–Scherrer procedure, an X-ray film is placed opposite a cylindrical powder preparation (adjusted in the center; crystal-powder method). A sample in the form of fine crystalline powder (particle size 0.1 mm) is irradiated by monochromatic X-ray light. In any case, the incident X-ray beam meets with a large number of tiny crystals, which satisfy Bragg's equation in

their position. Then the total of the reflected rays of a lattice series lies on the envelope of a cone with a particular aperture. At the point where the envelope of the cone cuts the film cylinder, the film turns black (further developed into the counter goniometer procedure with standard equipment as in XRFA; see 7.3).

C. Structural models. The steps involved in molecular structure analysis can be described in principle as follows. The grid parameters and the number of formula units in the unit cell are determined from the diffraction intensities, and a spatial group determination is made. The actual arrangement or selection of structure models starts with the evaluation of the Patterson function (mathematical way to calculate the approximate atomic position) and/or electron density function. This is followed by a refinement involving the completion of the geometrical (1) and thermal (2) parameters. The final representation of the structure takes into account the bond lengths and angles, symmetry, molecular packing, intermolecular interactions and the electron density distribution of the molecules.

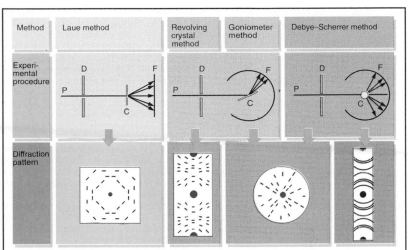

P: primary beam, D: diaphragm, C: crystal, F: film

B. X-ray diffraction methods

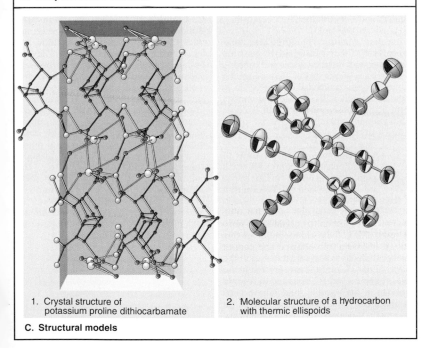

1. Crystal structure of
   potassium proline dithiocarbamate

2. Molecular structure of a hydrocarbon
   with thermic ellispoids

C. Structural models

11.6 Elemental Species Analysis

Introduction. Elemental species analysis, or speciation, deals with the analysis of physical and chemical forms of elements, especially of metals, in their matrix, be that air, water, soil or organisms. To be able to evaluate the mobility of heavy metals, bioavailability in plants, resorption behavior in animals and humans and toxic effects, differentiated knowledge above and beyond that of the total content is essential regarding the form of bonds of a metal in its matrix. Speciation provides the necessary analytical data for these questions.

A. Compartmentalization of lead and other heavy metals. Compartmentalization refers to the division of a sample into individual components, shown in the example by the components of a plant (feed grasses along a motorway). The analysis of element levels, lead in this case, in these compartments represent a differentiated analysis and a contribution to speciation.

1. Analysis diagram. The stored lead component is obtained (wash liquid) by washing the leaves with distilled water and rubbing them with a paper towel. Crushing the leaves on glass beads makes it possible to separate the cell walls. The cell organelles in the filtrate are separated from the vacuole fluid with the soluble proteins by ultracentrifugation.

2. Compartmentalization. About 64% of the total lead content (14 ppm) of a fresh rootless plant is found on the surface of the leaves, and can be washed off. The amount of lead inside the leaves is largely bound relatively tightly to the cell walls. It can be brought into solution (by about 80%) only by using dilute hydrochloric acid. In this way, it has been immobilized and cannot easily effect physiological processes in the plant. About 3% of the lead is present in the cell organelles and 5% is bound to soluble proteins in the vacuole fluid. These constituents can have damaging effects on enzymatic processes and membrane properties.

3. Mobility from soils. Extraction solutions with buffers having a different pH are used for studies on the mobility of heavy metals from soils. There is a standard extraction sequence which extends from an acetate buffer of pH 7 to the addition of strongly oxidizing acids. For example, the metals cadmium, zinc, copper and lead present very different pH-dependent mobilities. Acid rain effects include the release of heavy metals from soils.

B. Heavy metal species in natural waters. After ultrafiltration, the dissolved component of metals in waters can be divided into two groups analytically: that portion which can be determined directly without preprocessing the sample using voltammetry or polarography is called the electrochemical active element trace. The chemically bound components can only be detected after UV decomposition of the organic materials (after acidification and with the addition of H_2O_2).

C. Elemental species in foodstuffs. Bonding partners of metals, minerals and trace elements in foodstuffs include proteins, phenolic acids and pectins. They can be categorized according to their physical-chemical properties. Available separation methods include extraction procedures, electrophoresis, gel chromatography (especially for proteins), HPLC using different detectors (including post-column derivatization, see p. 160); available determination methods include spectral photometry and especially AAS. The questions come from food technology (e.g. discoloration due to the formation of metal complexes, crystallization of calcium salts in wine) and from nutritional science (in matters of resorption, such as of iron, and toxicology, as with respect to inorganic and organic mercury levels in fish).

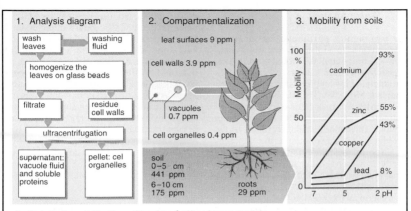

1. Analysis diagram

wash leaves → washing fluid

homogenize the leaves on glass beads

filtrate | residue cell walls

ultracentrifugation

supernatant: vacuole fluid and soluble proteins | pellet: cell organelles

2. Compartmentalization

leaf surfaces 9 ppm
cell walls 3.9 ppm
vacuoles 0.7 ppm
cell organelles 0.4 ppm

soil
0–5 cm 441 ppm
6–10 cm 175 ppm
roots 29 ppm

3. Mobility from soils

cadmium 93%
zinc 55%
copper 43%
lead 8%

Mobility %

pH   7   5   2

A. Compartmentalization of lead and other heavy metals

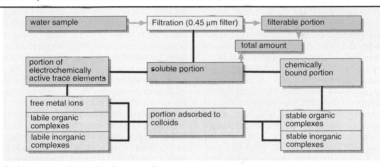

water sample → Filtration (0.45 μm filter) → filterable portion

total amount

portion of electrochemically active trace elements | soluble portion | chemically bound portion

free metal ions
labile organic complexes
labile inorganic complexes

portion adsorbed to colloids

stable organic complexes
stable inorganic complexes

B. Heavy metal species in natural waters

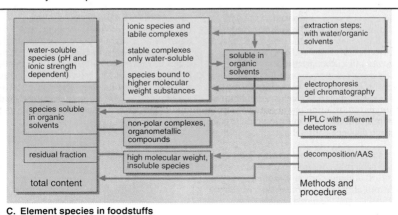

water-soluble species (pH and ionic strength dependent)

ionic species and labile complexes
stable complexes only water-soluble
species bound to higher molecular weight substances

soluble in organic solvents

extraction steps: with water/organic solvents

species soluble in organic solvents

non-polar complexes, organometallic compounds

electrophoresis gel chromatography

HPLC with different detectors

residual fraction

high molecular weight, insoluble species

decomposition/AAS

total content

Methods and procedures

C. Element species in foodstuffs

11.7 Water Analysis

A. Summation parameters and group parameters. In addition to the determination of individual dissolved substances, physical-chemical values and summation parameters play an important role in water analysis. Most often the goal is to determine the burden from organic substances. To determine the degree of burden from organic substances, with reference to the biological equilibrium and the autopurification of the particular body of water, methods were developed for a summary determination which are based on the chemical oxidizability of organic substances. The chemical oxygen demand (COD) is the amount of oxygen per liter required for the oxidation of organic substances to CO_2 and H_2O using potassium dichromate under defined conditions.

The biochemical oxygen demand (BOD) is defined as the volume-related mass of oxygen utilized by microorganisms at 20 °C to break down organic materials which are present. These processes require considerably more time than chemical oxidation: up to 20 days (BOD_{20}) with residential effluents. Usually the BOD_5 is determined when approximately 70% decomposition has been achieved.

The TOC and DOC, i.e. total and dissolved organic carbon, represent the proportion of organically bound carbon via the determination of CO_2. The TC, or total carbon content, is composed of the TIC, or total inorganic carbon content (carbonates and CO_2), and the TOC. The TOC in turn consists of the DOC and the POC (particulate organic carbon portion).

Organic halogen compounds in the form of AOX, EOX and POX are still determined as group parameters, as adsorbable, extractable or exhaustible halogen compounds, the determination of which is made via the halogen portions.

B. TOC functional principle. The analytical principle consists of the decomposition of organic carbon compounds using UV radiation after the addition of peroxodisulfate and the measurement of the CO_2 formed due to the absorption of IR radiation with a nondispersive IR detector (NDIR detector; see p. 210). After sample dosing, for liquid, solid or suspended samples, the TIC value can be determined by adding phosphoric acid, then the TOC value is determined after the addition of peroxodisulfate and oxidation in a UV reactor. High-temperature decomposition is necessary for solids and suspensions. The water which is produced or evaporated by oxidation in the condenser–water separator system is removed before the IR absorption measurement. A personal computer is used to control the device and the automatic analysis process, including the data processing.

C. Diagram of an AOX device. The analyte is burned in oxygen at 1000 °C and the combustion products proceed to the electrolysis cell, where they are titrated coulometrically (see p. 70). The apparatus consists of the combustion device (furnace) with gas supply and flow meters, a titration device with a drying component (concentrated sulfuric acid) and a titration cell with a coulometer. The coulometer generates the electrolyte flow, in this case for a microcoulometric titration (absolute method) of halide ions (Ag^+ ions) from the anode with subsequent potentiometric (or amperometric) induction. The cathode and anode are silver electrodes. The coulometer measures the amount of electricity required for the titration and converts the electricity into impulses for the digital display. Using this principle of measurement and the described apparatus, AOX, EOX and POX values can be determined in drinking water, process water and effluents (DIN 38409, Part 14), In the AOX procedure, the organic halogen compounds are adsorbed on activated charcoal, which is then burned in a quartz boat in the furnace.

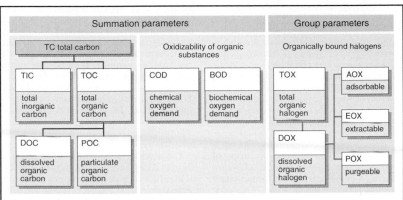

A. Summation and group parameters

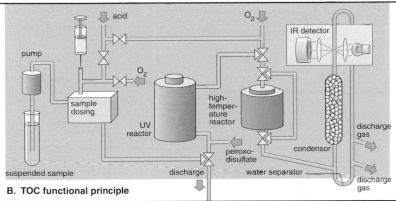

B. TOC functional principle

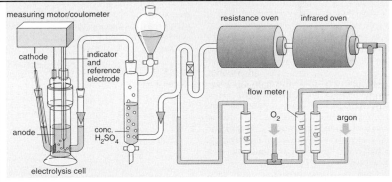

C. Diagram of an AOX device

11.8 Analysis of Aroma Substances

Introduction. The total sensory character, or flavor, of a food which is evoked when eaten occurs as a result of flavorings and odiferous compounds or aroma substances. The volatile components of a food having different chemical structures which induce a typical character or impact in the oronasal region are given the general label 'aroma compounds.'

A. Aroma impact compounds. The character of a food is determined by a number of substances. A requirement for a substance to have aromatic activity is a distinct vapor pressure at room temperature. The odor limits or threshold concentrations extend from 2 ppt (for 2-isobutyl-3-methoxypyrazine in potatoes) to 100 ppm for ethanol. The number of volatile compounds which have been identified extends from 59 in pineapple to over 600 in wine. The key compounds of an aroma are called 'character impact compounds,' and they come from the groups of aliphatic hydrocarbons, alcohols, esters, aldehydes, ketones, ethers, lactones and thiazoles. In some cases, the aroma is determined by just a few impact compounds, for example by oct-1-en-3-ol in mushrooms, by several aldehydes such as hexanal in tomatoes or by ketones in raspberries. A larger number of compounds are needed to come close to simulating the aroma of a peach, for example (lactones, different esters, alcohols, acids and benzaldehyde). Incorrect aromas can develop due of the loss of impact compounds which are formed in foods as a result of natural processes, or due to foreign aroma substances or changes in the concentration ratios of the individual aroma substances.

B. Methods and procedures in the analysis of aroma substances. Composite procedures derived from separation and determination methods are necessary in the analysis of aroma substances. This analysis plays an important part in food technology, for the evaluation of raw materials and products and in processing.

Separation techniques that are used include distillation (from aqueous–liquid foods), extraction (with organic solvents or with pressurized CO_2), simultaneously in an apparatus as per Lickens and Nickerson, the headspace technique (see p. 170)—in connection with adsorption methods or with gas chromatography. Polymeric substances such as Porapak (base is divinylbenzene–styrene or ethylvinylbenzene) or Tenax (polyphenylene oxide) are used as adsorption materials. Preseparations using preparative gas chromatography or classical column chromatography on silica gel follow after a concentration step.

The aroma fractions thus obtained are separated gas chromatographically, usually in thin-film capillaries in conjunction with mass spectrometric detection (for identification), either directly or after another concentration step, and possibly after derivatizing functional groups. HPLC is rarely used for this. If the mass spectra do not provide unambiguous information, additional molecular spectroscopic methods are used, such as ^{1}H-NMR or IR (also a Fourier transform IR detector for GC). Sensory analysis (sensory–physiological taste and odor analysis based on objective criteria) is also of great importance in the framework of the analysis of aroma substances, e.g. the 'sniffing test' in conjunction with gas chromatography ('the human nose as a detector'). By smelling the carrier gas stream, one obtains an 'aromagram,' i.e. individual fractions are assigned to specific odor (aroma) impacts, e.g. roasting flavors in the aromagram for coffee. A structure for an isolated aromatic substance which is based on spectroscopic information is identified unambiguously if it agrees with a synthesized reference substance ('chemical synthesis').

Group	Name	Occurrence	Structure
hydrocarbon	1,3,8-p-menthatriene	parsley	
alcohol	menthol	peppermint	
ester	2-trans-4-cis-deca-dienoic acid ethyl ester	pear	
aldehyde	2-trans-6-cis-nonadienal	cucumber, chopped	
ketone	1-(4-hydroxyphenyl)-butan-3-one	raspberry	
ether	1,8-cineol	eucalyptus	
lactone	3-isobutylidene-3,4-dihydrophthalide	celery	
thiazole	2,4-dimethyl-5-vinylthiazole	nuts	

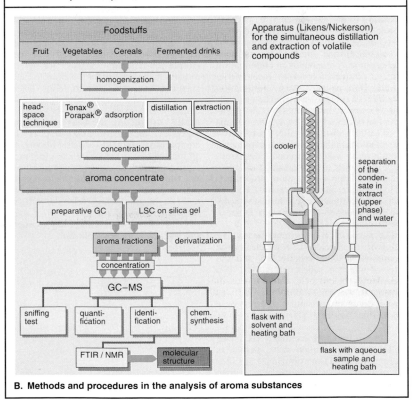

A. Aroma impact compounds

B. Methods and procedures in the analysis of aroma substances

11.9 Pesticide or Residue Analysis

Introduction. Pesticide or residue analysis fundamentally includes analytical procedures for trace elements. The sample preparation, or reprocessing, represents a special focal point, since it must be coordinated with the sample type (food obtained from plants or animals) and with the selectivity of the determination methods. Primarily chromatographic procedures with special detection methods are considered for this type of analysis.

A. Work flow for residue analysis of plant materials (according to Thier and Frehse). The first step of any residue analysis consists of the extraction of the experimental material, e.g. with acetone or acetonitrile. Acetonitrile extracts are used according to the guidelines of the AOAC (Association of Official Analytical Chemists) in the USA for the determination of 26 organochlorine compounds and organophosphorus compounds in combination with a distribution step (acetonitrile + water/light petroleum) and purification on a Florisil (magnesium silicate gel) column. However, only slightly polar pesticides can be detected this way. One obtains more polar substances after distribution with dichloromethane and purification on a column of activated charcoal, magnesium oxide and Celite (diatomaceous earth mixture). Available purification procedures include column chromatography (CC) with different adsorbents or gel chromatography (GC) and sweep co-distillation (SCD; see p. 30).

The purification steps must be coordinated with the efficiency, i.e. selectivity, the separation method, usually GC, and the detector which is subsequently used. If a phosphorus-selective GC detector is used, considerably simpler reprocessing suffices than if an electron-capture detector or a thermoionic detector (TID) is used.

B. Adsorption and capillary GC with ECD. After distribution steps, pesticides such as atrazine (1), metribuzin (2), alachlor (3), metolachlor (4) and cyanazine (5) can be adsorbed from an aqueous phase, even on a reversed-phase (C_{18}) material, which has particularly small blank values. The subsequent GC analysis took place in a fused-silica capillary column 30 m in length coated with silicone OV-1701 and an ECD as the detector (2).

C. GC with mass-selective detection. Organophosphorus compounds can be detected with a mass-selective detector in addition to a phosphorus-selective detector. This means that the chromatogram can be recorded by measuring the total ion current (TIC). The high efficiency of capillary GC makes possible the complete separation of 29 pesticides in about 40 min. A mass spectrometer attached to a gas chromatograph can take on more functions in addition to that of a non-specific detector: complete mass spectra of especially interesting peaks or eluted substances can be recorded within a few seconds. A mass-selective detector is one for which the gas chromatograms are registered for a firmly established mass (single-ion detection: SID) or for several masses characteristic for the expected substance concurrently (multiple-ion detection: MID). Data systems make it possible to display 'corrected' spectra—by electronically subtracting the background—and especially the identification of peaks by comparing them with a spectral library. An additional role is played by mass fragmentography, in which measurements are made via time-programmed individual recordings or via multiple mass registrations (MMR). In spite of high selectivity, in this instrumental pesticide analysis one must pay attention to interference due to impurities which were not separated.

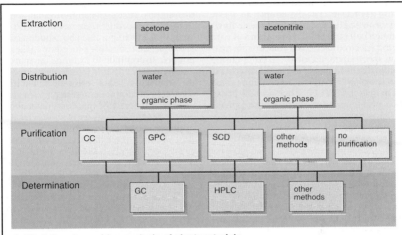

Extraction
acetone | acetonitrile

Distribution
water / organic phase | water / organic phase

Purification
CC | GPC | SCD | other methods | no purification

Determination
GC | HPLC | other methods

A. Work flow for residue analysis of plant materials

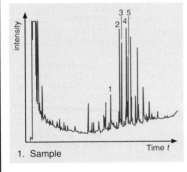

1. Sample

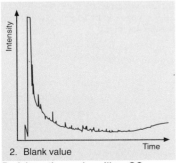

2. Blank value

B. Adsorption and capillary GC with ECD

3. Pesticides

atrazine (1)

alachlor (3)

metribuzine (2)

metolachlor (4)

cyanazine (5)

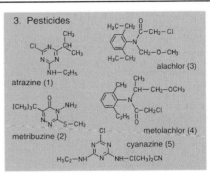

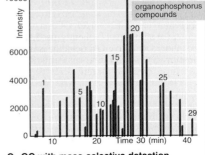

organophosphorus compounds

C. GC with mass-selective detection

D. Analysis of carbamate pesticides using HPLC–FTIR. To couple an FTIR device to an HPLC device, a suitable flow-through cell is inserted into the path of rays of the spectrometer. As a rule, a transmission spectrum is recorded. To compensate for the IR absorption of the mobile phase, its spectrum is recorded beforehand. This technique is well suited for normal-phase LC; problems arise with reversed-phase systems with water as a strongly IR-absorbing solvent.

The carbamates carbaryl and chlorpropham are usually separated on silica gel with 1,2-dichloroethane as the mobile phase within the framework of pesticide residue analysis. The detection is done at the peak maximum of the CO band region using a CaF$_2$ cuvette. The results can be verified (checked for identity and peak purity) just from the section of the IR spectrum shown, which is recorded during the chromatographic separation and then is printed out later.

E. Atrazine analysis with a photodiode array detector. HPLC is better suited for the determination of the polar triazine and phenylureas than is GC. According to a DIN procedure (38407 F12), 17 substances and more, as well as metabolites, can be detected, the triazines being detected at 230 nm, the phenylureas at 245 nm and some aniline derivatives at 214 nm. The peak purity check and the identification take place as follows: based on the retention time, comparisons with reference samples can lead to the supposition that the example shown of a ground water sample is atrazine (1) at a detection wavelength of 230 nm. The spectrum of an atrazine standard is stored in the spectral library (2). Superposition of the spectra (3) shows that both spectra, that of the standard and the substance eluted with a retention time of 23.2 min, are identical.

F. Triazine analysis with GC and an atomic emission detector. Coupling capillary or high-resolution gas chromatography with an atomic emission detector (AED, see p. 92) makes it possible to detect specific elements such as carbon, nitrogen, chlorine or sulfur. The results of coupling GC and AED are element-selective chromatograms. Just as when molecular spectrometric procedures are used in HPLC analysis according to the photodiode-array principle, this coupling procedure makes it possible to continue to examine the individual peaks for possible interferences due to the matrix or other substances via the recording of atomic emission partial spectra. The efficiency of this coupling technique in pesticide analysis is shown by the example of the examination of a surface water extract: the capillary gas chromatogram with carbon detection (similar to detection with an FID, but considerably more sensitive with the AED) would not make it possible to determine with certainty the pesticide sought for, i.e. ametryne (1). All vaporizable organic substances are displayed. On the other hand, if a sulfur-selective detection is performed, a well ordered chromatogram is obtained in which the signals 1–5 can be assigned to the sulfur-containing pesticides aziprotryne, ametryne, terbutryne, chlorthion and methoprotyn (2). Also, as in UV detection, the use of the AED as a photodiode-array detector makes it possible to detect a section of the emission spectrum (3)—possible for the range from 160 to 800 nm—with the emission lines at 180.7 to 182.6 nm, which are characteristic for sulfur. False-positive peaks, produced by matrix effects or interfering substances, can be detected in this way: all three lines must be present and their intensities must continue to show a specific ratio. N- and Cl-selective detection can be performed in the same way. Because of the detection principle, element sensitivities in the pg/s range can be achieved with atomic emission detectors.

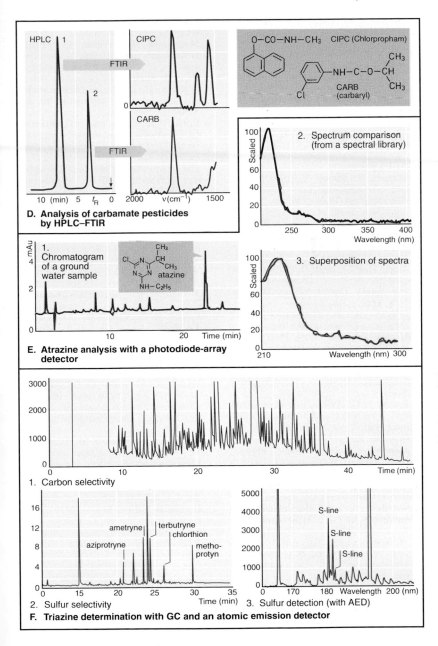

D. Analysis of carbamate pesticides by HPLC–FTIR

2. Spectrum comparison (from a spectral library)

3. Superposition of spectra

E. Atrazine analysis with a photodiode-array detector

1. Chromatogram of a ground water sample

1. Carbon selectivity

2. Sulfur selectivity

ametryne | terbutryne
aziprotryne | chlorthion
metho-protyn

3. Sulfur detection (with AED)

S-line

F. Triazine determination with GC and an atomic emission detector

Bibliography

Figures from the following books served as models:

- Fifield, F. W., Kealey, D. (1990): *Principles and Practice of Analytical Chemistry*, 3rd edn, Blackie, Glasgow/London.
- Metzner, K. (1977): *Gaschromatographische Spurenanalyse*, Leipzig.
- Schröder, B., Rudolph, J. (Eds) (1985): *Physikalische Methoden in der Chemie*, VCH, Weinheim.
- Wagner, H., Blasius, E. (1989): *Praxis der elektrophoretischen Trennmethoden*, Springer-Verlag, Berlin, Heidelberg, New York.
- Oehme, E., Jola, M. (1982): *Betriebsmesstechnik unter Einsatz von in-line und on-line Analysatoren*, Springer-Verlag, Heidelberg.
- Thier, H.-P., Frehse, H. (1986): *Rückstandsanalytik von Pflanzenschutzmitteln*, Georg Thieme Verlag, Stuttgart, New York.
- Williams, D. H., Fleming, I. (1991): *Strukturaufklärung in der organischen Chemie*, 6th edn, Georg Thieme Verlag, Stuttgart, New York.
- Strobel, H. A., Heineman, W. R. (1988): *Chemical Instrumentation: a Systematic Approach*, 3rd edn, New York.
- Hargis, L. G. (1988): *Analytical Chemistry, Principles and Techniques*, Prentice Hall, Englewood Cliffs, NJ.
- Geckeler, K. E. (1987): *Analytische und präparative Labormethoden*, Vieweg, Braunschweig, Wiesbaden.
- Henze, G., Neeb, R. (1986): *Elektrochemische Analytik*, Springer-Verlag, Berlin, Heidelberg.
- Rump, H. H., Krist, H. (1987): *Laborhandbuch für die Untersuchung von Wasser, Abwasser und Boden*, VCH, Weinheim.
- Günzler, H., Böck, H. (1983): *IR-Spektroskopie. Eine Einführung*, 2nd edn, VCH, Weinheim.
- Hahn-Weinheimer, P., Hirner, A., Weber-Diefenbach, K. (1984): *Grundlagen und praktische Anwendung der Röntgenfluoreszenzanalyse (RFA)*, Braunschweig.
- Mannkopff, R., Friede, G. (1975): *Grundlagen und Methoden der chemischen Emissionsspektralanalyse*, VCH, Weinheim.
- Günther, H. (1983): *NMR-Spektroskopie*, 2nd edn, Georg Thieme Verlag, Stuttgart, New York.
- Rücker, G., Neugebauer, M., Willems, G. G. (1988): *Instrumentelle pharmazeutische Analytik, Lehrbuch zu spektroskopischen, chromatographischen und elektrochemischen Analysenmethoden*, Wissenschaftliche Verlagsgesellschaft, Stuttgart.
- Westermeier, R. (1990): *Elektrophorese-Praktikum*, VCH, Weinheim.
- Danzer, K., Than, E., Molch, D., Küchler, L. (1987): *Analytik. Systematischer Überblick*, Wissenschaftliche Verlagsgesellschaft, Stuttgart.
- Willard, H. H., Merritt, L. L., Dean, J. A., Settle, F. A. (1981): *Instrumental Methods of Analysis*, 6th edn., Belmont, CA.
- Pecsok, R. L., Shields, L. D.D, Caims, T., McWilliams, I. G. (1976): *Modern Methods of Chemical Analysis*, 2nd edn, Wiley, New York.
- Christian, G. D. (1980): *Analytical Chemistry*, 3rd edn, Wiley, New York.
- Schenk, G. H., Hahn, R. B., Hartkopf, A. V. (1977): *Quantitative Analytical Chemistry, Principles and Life Science Applications*, Houghton Mifflin, Boston, MA.
- Peters, D. G., Hayes, J. M., Hieftje, G. M. (1974): *Chemical Separations and Measurements*, Saunders, Philadelphia, PA.
- Bauer, H. H., Christian, G. D., O'Reilly, J. E. (1978): *Instrumental Analysis*, Houghton Mifflin, Boston, MA.
- Riesen, R., Widmann, G. (1984): *Thermoanalyse, Anwendungen, Begriffe, Methoden*, Springer-Verlag, Heidelberg.
- Budzikiewicz, H. (1980): *Massenspektrometrie*, 2nd edn, VCH, Weinheim.
- Doerffel, K., Eckschlager, K. (1981): *Optimale Strategien in der Analytik*, Thun, Frankfurt.

- Schomburg, G. (1987): *Gaschromatographie, Grundlagen, Praxis, Kapillartechnik*, 2nd edn, VCH, Weinheim.
- Doerffel, K., Müller, H., Uhlmann, M. (1986): *Prozessanalytik*, Leipzig.
- Naumer, H., Heller, W. (Eds) (1990): *Untersuchungsmethoden in der Chemie, Einführung in die moderne Analytik*, 2nd edn, Georg Thieme Verlag, Stuttgart, New York.
- Analytikum (1987) *Methoden der analytischen Chemie und ihre theoretischen Grundlagen*, 7th edn, Leipzig.
- Grasserbauer, M., Dudek, H. J., Ebel, M. F. (1985): *Angewandte Oberflächenanalyse mit SIMS Sekundär-Ionen-Massenspektrometrie, AES Auger-Elektronen-Spektrometrie XPS Röntgen-Photoelektronen-Spektrometrie*, Springer-Verlag, Berlin, Heidelberg, New York, Tokyo.
- Oster, H. (1973): *Prozess-Chromatographie*, Frankfurt.
- Günther, W., Schlegelmilch, F. (1984): *Gaschromatographie mit Kapillar-Trennsäulen, Grundlagen*, Würzburg.
- Welz, B. (1983): *Atomabsorptionsspektrometrie*, 3rd edn, VCH, Weinheim.

- Bock, R. (1974, 1980, 1984): *Methoden der Analytischen Chemie. Eine Einführung. Volume 1: Trennungsmethoden, Volume 2: Nachweis- und Bestimmungsmethoden, Parts 1 and 2*, VCH, Weinheim.
- Fritz, J. S., Schenk, G. H. (1989): *Quantitative Analyse*, Vieweg, Braunschweig, Wiesbaden.
- Ruzicka, J., Hansen, E. H. (1981): *Flow Injection Analysis*, Wiley, New York.
- Cammann, K. (1987): *Das Arbeiten mit ionenselektiven Elektroden*, 3rd edn, Springer-Verlag, Heidelberg, Berlin, New York, Tokyo.
- Schwedt, G. (1986): *Chromatographische Trennmethoden*, 2nd edn, Georg Thieme Verlag, Stuttgart, New York.
- Schwedt, G. (1980): *Fluorometrische Analyse*, VCH, Weinheim.
- Kunze, U. R. (1990): *Grundlagen der quantitativen Analyse*, 3rd edn, Georg Thieme Verlag, Stuttgart, New York.
- Schwedt, G. (1995): *Analytische Chemie. Grundlagen, Methoden und Praxis*, Georg Thieme Verlag, Stuttgart, New York.

Index

Absolute mass range, 10
Absorption spectrometry, 104, 105, 108, 109
  applications, 108, 109, 110, 111
  phototometers used, 108, 109
  see also UV–VIS spectrophotometry
Acid–base titrations, 40, 41
  indicators used, 40, 41
  potentiometric determination of end point, 60, 61
Acidimetry, 40, 41
Additive errors, 16
Adrenaline, determination of, 188, 189
Adrenochromes, 188, 189
Adsorption, separation by, 140, 141, 146
Adsorption isotherms, 146, 147
Adsorption thin-layer chromatography, 148, 149, 196
  comparison with HPLC, 152, 153
Aerobacter aerogenes 78, 79
Air analysis
  detector tubes for, 36, 37
  sampling methods for, 18, 19
Alkali-metal sulfates, thermal differentiation of, 78, 79
Alkali metals, detection/separation of, 34, 35
Alkalimetry, 40, 41
Alkaline-earth metal sulfates, thermal differentiation of, 78, 79
Alkaline-earth metals, detection/separation of, 34, 35
Alkanes
  GC retention times, 166, 167
  IR spectra, 118, 119
Alkylbenzenes MS spectra, 128, 129
Aluminum
  determination of, 102, 103
  wet-chemical detection/separation of, 33, 38
Ammonia optosensor, 208, 209
Ammonium, determination in water, 192, 193

Amperometric detectors, 158, 159, 192
Analytical chemistry *see* Chemical analysis
Analytical method, relationship to principle and procedure, 6, 7
Analytical methods, 8
  application fields/ranges, 10, 11
  classification of, 6, 7
  comparison of, 10, 11
  symbols for, 9
Analytical principle, 6, 7
Analytical procedure, 6, 7
Analytical strategy, 4, 5
Animal proteins, analysis of origin, 180, 181
Anthracene, 113
  UV spectra, 110, 111
Antibody–atrazine reaction, 48, 49
Antigen–antibody reaction, 48, 49, 180, 181
Antigorite, 78, 79
Antimony
  determination of, 84
  separation of, 33, 34
Apple aroma, GC analysis of, 172, 173
Argentimetric precipitation titrations, 42, 43
  see also Silver...
Aroma compounds
  analysis of, 172, 173, 224–5
  character impact compounds, 224, 225
Aromagrams, 224
Arsenic
  determination of, 102, 103
  separation of, 32, 33, 34
Atomic absorption spectrometry (AAS), 6, 82–91
  applications, 86, 87, 90–1, 220
  atomizer/mixing chamber burner used, 84, 85, 90
  background correction, 88, 89
  chemical interference suppression, 86, 87

chemicophysical processes in flame, 82, 83
coupled with liquid chromatography, 198, 199
deuterium background corrector, 88, 89
device with graphite furnace, 82, 83
graphite tube furnace, 84, 85
    absorption signals when heated, 86, 87
    temperature program for, 86, 87
hollow-cathode lamp, 84, 85
hydride technique, 84, 85, 90, 91
influence of spectral bandwidth, 86, 87
ionization interference suppression, 86, 87
laser spectrometry and, 204, 205
measuring principle, 82, 83
spectral interference suppression, 86, 87
spectrochemical buffer used, 86, 87
Zeeman correction, 88, 89
Atomic emission spectrometry (AES), 6, 92–7
coupled with GC, 196, 197, 228, 229
excitation sources used, 92, 93
flame excitation, 94–5
plasma excitation, 96–7
ray path in spectrograph, 92, 93
Atomic spectrometric methods, 6, 82–103
concentration range, 10, 11
Atrazine
determination of, 48, 49, 226, 227, 228, 229
molecular structure, 49, 227
Auger effect, 98, 99, 202
Auger electron spectrometry (AES), 202, 203
Automation of analytical procedures, 18, 19, 24, 25, 30, 184–99
see also Continuous flow analysis; Coupling techniques; Flow injection analysis
Averages, calculation of, 14, 15

Balmer series [for emission spectrum of hydrogen], 92, 93
Barium
alkaline error of pH glass electrodes, 55
determination of, 102, 103
separation of, 33, 34, 35, 38

Bathochromic effect, 110
Benzene and derivatives
IR spectra, 118, 119
and nuclear resonance, 134, 135
UV spectra, 108, 109, 110, 111
Beta scintillation counters, 200, 201
Bibliography, 230–1
Bioaffinity chromatography, 162, 163
Biochemical oxygen demand (BOD), 222
determination in sewage water, 212
Biosensors, 56
see also Chemical sensors
Bismuth
determination of, 84
separation of, 33, 34, 35
Bladder stones, thermal analysis of, 78, 79
Blank values, 14, 15
Blood analysis, 90, 91, 170
Bouger–Lambert–Beer law, 106
see also Lambert–Beer law
Bowed calibration curves, 16, 17
Bragg's law, 100
Brunauer–Emmet–Teller (BET) adsorption isotherm, 146, 147
Burettes, 40, 41
n-Butane, NMR spectra, 136, 137
n-Butylbenzene, MS spectra, 128, 129

Cadmium
determination of, 64, 68, 69, 86, 87
mobility from soils, 220, 221
separation of, 33, 34, 35
Cadmium sulfate, decomposition voltage, 53
Calcium
separation of, 33, 34, 35
titrimetric determination of, 42, 43
Calcium-selective electrode, calibration curve for, 58, 59
Calibration curves, 16–17
chromatographic methods, 152, 153
falsification of, 16, 17
fluorimetry, 16, 17, 112, 113
ion-selective electrodes, 58, 59, 194, 195
photometry, 16, 17, 110, 111
standard addition method used, 16, 17
Calomel electrodes, 56, 57

Calorimetry, 72, 73
    differential scanning calorimetry, 72, 73, 80, 81
Capacity voltage, compared with Faraday voltage, 66, 67
Capillary electrophoresis, 182, 183
Capillary gas chromatography, 166, 167
    coupled with HPLC, 196, 197
Capillary liquid chromatography, 156, 157
    coupled techniques
        with GC, 196, 197
        with MS, 196, 197
        with TLC, 196, 197
    see also High-performance liquid chromatography
Carbamate pesticides, analysis of, 228, 229
Carbonyl compounds, α-cleavage of, 130
Carbonyl group, molecular vibrations, 104, 105, 117
Cartridges, enrichment using, 28, 29
Catecholamines, determination of, 188, 189
Cathodic preconcentration, 28
Cation analysis
    polarographic methods, 64, 65, 68, 69
    wet-chemical methods, 32, 33
Cellulose acetate strips, in carrier electrophoresis, 176, 177
Cement industry, process analysis in, 214, 215
Centrifugation
    separation by, 8, 140, 141
    symbol for, 9
Chemical analysis, basic principles, 2–17
Chemical oxygen demand (COD), 222, 223
Chemical reaction detector [HPLC], 160, 161
Chemical sensitive field effect transistor (CHEMFET), structure, 208, 209
Chemical sensors, 206–9
    see also Biosensors
Chemisorption sensors, principles, 206, 207
Chlorinated phenols, GC analysis of, 172, 173
Chlorohydrocarbons, IR spectra, 116, 117, 120, 121

2-(4-Chlorophenoxy)-2-methylethyl propionate, NMR spectra, 138, 139
Chromate–sulfate [separation] procedure, 34, 35
Chromatography, 6, 142–73
    concentration range, 10, 11
    dynamic theory, 144, 145
    elution curve, 142, 143
    fronting of peaks, 144, 146
    kinetic theory, 142, 143
    linear non-ideal behavior, 144, 145
    non-linear ideal behavior, 144, 145
    resolution of bands/peaks, 142, 143
    sample preparation using, 30, 31
    separation mechanisms, 146–7
    symbols for, 9
    tailing of peaks, 144, 146
    theoretical plate model, 142, 143
    two-dimensional, 196
    see also Bioaffinity...; Gas...; Gel permeation...; High-performance liquid...; Ion-exchange...; Liquid...; Reversed-phase...; Thin–layer chromatography
Chromium
    determination of, 102, 103
    wet-chemical detection/separation of, 32, 33
Chromium sulfate, decomposition voltage, 53
Chromophores, 108
Clay minerals, thermal analysis of, 78, 79
Cleanup procedures, 30–1
^{13}C-NMR spectroscopy, 138, 139
Co-distillation see Sweep co-distillation
Co-precipitation, enrichment by, 28, 29, 140
Cobalt, wet-chemical detection/separation of, 33, 36, 37
Cobalt sulfate, decomposition voltage, 53
Coenzymes, 44
Coffee extracts/aroma, analysis of, 200, 201, 224
Cold plasma incineration, 26
Colorimetry, 106–7
    devices
        immersion colorimeter, 106, 107
        sliding comparator, 106, 107
        turntable comparator, 106, 107
    principles, 106

Column liquid chromatography, 154–63
  coupled with AAS, 198, 199
  see also Liquid chromatography
Column-switching techniques, 30, 31, 170, 171
Combustion *see* Incineration methods
Comparators, colorimetry, 106, 107
Complexing agents, 34, 35
Complexometric titrations, 40, 42, 43
Composite methods, 12, 13
  sources of error, 12, 13
Concentration ranges, 10, 11
  nomenclature used, 10, 11
Condensation, separation by, 8, 140, 141
Conductometry, 60–1
  measuring cells used, 60, 61
  principles of measurement, 60, 61
Confidence interval, 14
Content analysis, 2, 3
Continuous flow analysis (CFA), 184, 186–9
  air-segmented liquid flow, 186, 187
  applications, 188–9
  dispersion dependences, 186, 187
  signal forms, 186, 187
  structure of system, 186, 187
Copper
  determination of, 64, 68, 69, 96, 97, 102, 103
    calibration curve for, 16, 17
  mobility from soils, 220, 221
  photoelectron spectra, 204, 205
  wet-chemical detection/separation of, 33, 34, 35, 37
Copper group elements, separation of, 33, 34, 35
Copper sulfate
  decomposition voltage, 53
  electrolysis of, 52, 53
Corynebacterium diphtheria, 78, 79
Coulometric titration, 70, 71
Coulometry, 70–1
Coumarin, 113
Counter tube [for X-ray fluorescence], 100, 101
Coupling techniques, 196–9
Craig countercurrent apparatus, 154
Crystallization
  separation by, 140, 141
  symbol for, 9

Current–time curves
  coulometry, 70, 71
  polarography, 66, 67
Current–voltage curves
  electrolysis, 52, 53
  polarography, 64, 65
  voltammetry, 62, 63
  see also Voltage–current curves
Cuvettes
  colorimetry, 106, 107
  IR spectrometry, 114, 115
  spectrophotometry, 108, 109
Cyanide, determination of, 188, 189
Cyclohexane, MS spectra, 130, 131
Cyclohexanone, and nuclear resonance, 134, 135
Cyclohexene, MS spectra, 130, 131
Cylinder pore model [in gel permeation], 162, 163

Dead-stop process, 70
Deans switch [in GC], 170, 171
Debye–Scherrer procedure [for powder X-ray diffraction], 218, 219
Decomposition methods [for sample preparation], 22–7, 222
Decomposition temperatures
  oxalates, 74, 75
  polymers, 74, 75
Decomposition voltages [in electrolysis], 50, 52, 53
Dehydrogenases, 44, 46, 47
Densitometric analysis, 152, 153, 176, 177
Deoxyribonucleic acid (DNA), analysis of, 182, 183
Deoxyribonucleosides, characterization by thermal analysis, 80, 81
Derivative thermogravimetry (DTG), 72, 73, 74
Detection limits
  in automated analysis, 188
  ICP-MS, 198
  meaning of term, 14, 15, 32
  polarography, 68
  sodium-selective electrode, 58
  test papers, 36
Detection methods, 32–7, 70
Detectors
  gas chromatography, 170, 171, 172–3

liquid chromatography, 156, 157, 158, 159, 160, 161
X-ray fluorescence, 100, 101
Determination limit, 14, 15
Diacetyl oxime, 34, 35, 38
Dialysis cell, sample cleanup by, 194, 195
Dichloromethane, comparison of IR and Raman spectra, 120, 121
Diels–Alder cycloaddition reaction, 130
Diethyl ether, NMR spectra, 136, 137
Differential pulse anodic stripping voltammetry (DPASV), 68, 69
Differential pulse polarography, 62, 66, 67
    application to organic compounds, 68, 69
Differential scanning calorimetry (DSC), 72, 73, 80–1
    applications, 80, 81
    calorimeter used, 80, 81
    curves, 80, 81
Differential thermal analysis (DTA), 72, 73, 76–9
    applications, 78, 79
    characteristic signal forms, 76, 77
    compared with normal thermal analysis, 76, 77
    measuring cell, 76, 77
    sample mount, 76, 77
Digestion methods [for sample preparation], 22–7
    coupled with FIA technique, 192, 193
DIN regulations, symbols used, 8, 9
α, α′-Dipyridyl, 34, 35, 36
Direct current plasma (DCP) emission spectrometry, 96, 97
Direct procedures, 12, 13
Dispersion [band broadening]
    in chromatography, 144, 145
    in continuous flow analysis, 186, 187
    in flow injection analysis, 190, 191
Distillation
    separation by, 8, 140, 141
    symbol for, 9
    see also Sweep co-distillation
Distribution analysis, 2, 3, 126, 202
Distribution isotherms, 146, 147
Division of sample, 20, 21
Double-focusing mass spectrometer, 126, 127

in SIMS, 126, 127, 202, 203
Drinking water analysis, 68, 69
Dropping mercury electrode (DME), 62, 63, 64
Dubosq's immersion colorimeter, 106, 107
Dust analysis, 102, 103
Dynamic analysis, 2, 3
Dynamic differential calorimetry (DDC), 80
    see also Differential scanning calorimetry
Dynamic ion-exchange model, 146, 147

Electroanalytical methods, 6, 7, 50–71
    concentration range, 10, 11
Electrochemical active element trace, 220
Electrochemical detectors, 158, 159, 192
Electrochemical double layer, 50, 51, 62
Electrodes
    pH-measurement, 54, 55
    polarography, 62, 63
    potentiometry, 56–7
Electrogravimetry, 52–3, 140, 141
    electrolytic device for, 52, 53
Electrolysis, 7, 50, 51
    overvoltage for, 50, 52
    sample enrichment using, 28, 29
    separation by, 8, 140, 141
    symbol for, 9
Electromagnetic spectrum, 104, 105
Electron capture detector (ECD), 172, 173, 226
Electron excitation, 104, 105
Electron impact ionization [in MS], 122, 123
Electron spectroscopy for chemical analysis (ESCA), 204
Electron-volt, energy value of, 122
Electrophoresis, 174–83
    applications, 176, 177, 178, 179, 180, 181, 182, 183, 220
    capillary electrophoresis, 182–3
    carrier electrophoresis
        applications, 176, 177
        techniques for, 174, 175, 176, 177
    carrier-free electrophoresis
        principles, 174, 175
        techniques for, 174, 175
    column electrophoresis, 176, 177

continuous carrier electrophoresis, 176, 177
countercurrent electrophoresis, 180, 181
disk electrophoresis, 178, 179
flow-through electrophoresis, 176, 177
isoelectric focusing, 174, 175, 180, 181
separation by, 140, 141, 174, 175
zone electrophoresis, 174, 175
  with immunodiffusion, 180, 181
  principles, 174, 175
Elemental species analysis, 200, 220–1
Elution
  in liquid chromatography, 154, 155
  in thin-layer chromatography, 148, 149
Elution chromatogram, 142, 143
Emission gas thermoanalysis (EGT), 72, 73
Emission line spectra
  hydrogen, 92, 93
  lithium, 94, 95
Emission spectrometry
  flame excitation used, 94–5
  optical emission spectrometry, 92–7
  plasma excitation used, 96–7
  X-ray emission spectrometry, 92
  see also Atomic emission spectrometry
Emission spectrum, 104, 105
  see also Fluorimetry
Empirical formula, analyses leading to, 216
End-point methods
  acid–base titration, 40, 41
  coulometric titration, 70, 71
  enzymic determination, 44, 45
Endothermic reactions, thermal analysis of, 72, 77, 80, 81
Energy-dispersive X-ray fluorescence analysis (EDXRFA), 100, 101
  applications, 102, 103
Energy-level/state diagrams
  see Term diagrams
Enrichment methods, 28–9
Enthalpimetry, 72, 73
Enzyme electrodes, 56, 57
Enzyme–substrate reactions, 44, 45
  reaction rate calculations, 44, 45
Enzyme-linked immunosorbent assay (ELISA) procedure, 48, 49
Enzymic analysis, 44–7
Ephedrine, MS spectra, 127

Epsomite, 78, 79
Errors
  sources
    calibration curves and, 16
    in composite methods, 12, 13
    in pH measurement, 54, 55
    in precipitation analysis, 38
    in sampling and sample preparation, 18, 19, 20, 22
    in spectroscopic methods, 102, 108
  see also Random...; Systematic errors
Escherichia coli, 78, 79
Ethyl groups, nuclear resonance lines, 136, 137
Ethylenediaminetetraacetate (EDTA), titrations involving, 42, 43, 58, 59
Exclusion volume [in gel permeation chromatography], 162
Exothermic reactions, thermal analysis of, 72, 77, 80, 81
Extraction, 140, 141
  for elemental species analysis, 220, 221
  symbol for, 9
  see also Liquid–liquid...; Solid–liquid extraction

Faraday current, 64, 66
Faraday voltage, compared with capacity voltage, 66, 67
Fertilizers, analysis of, 184
Field ionization laser spectrometry (FILS), 204
Filtration
  separation by, 8, 38, 39, 140
  symbol for, 9
Flame atomic emission spectrometry, 94, 95
  atomizer designs, 94, 95
  structure of photometer, 94, 95
Flame ionization detector (FID), 172, 173
Flame photometry, 6, 32, 94
Flow injection analysis (FIA), 184, 190–5
  applications, 192–5, 212, 213
  coupled with
    digestion procedures, 192, 193
    liquid–liquid extraction, 194, 195
  dispersion control, 190, 191
  optical sensors used, 208
  principles, 190, 191
  signal forms, 190, 191

dependence on sample volume, 190, 191

stopped-flow technique, 190, 191

Fluorescein, 113

Fluorescence
  indicators, 152, 153, 160, 161
  laser-induced, 204, 205
  three-dimensional diagram, 112, 113

Fluoride-selective electrode, 58, 194
  calibration curve for, 58, 59, 194, 195

Fluorimetry, 112–13
  automated system, 188, 189
  calibration curves, 16, 17, 112, 113
  concentration range, 10, 11
  structure of fluorimeter, 112, 113
  three-dimensional spectrum, 112, 113

Fluorinated hydrocarbons
  in IR spectrometry, 116
  spin–spin coupling in NMR, 134, 135

Fluorophores, 112, 113, 188

Foodstuffs, analysis of, 180, 181, 220, 221

Fourier transform infrared (FTIR) spectrometry, 118, 119
  coupled with HPLC, 228, 229

Freiberg decomposition, 32

Freundlich adsorption isotherm, 146, 147

Fructose, enzymic analysis of, 46, 47

Fumaric acid, polarographic determination of, 68, 69

Galvanic cell, 50, 51

Galvanic potential, 50, 51

Galvanostatic coulometry, 70, 71

Gas chromatography (GC), 6, 8, 164–73
  applications, 172, 173, 226, 227
  aroma compounds characterized using, 170, 171, 224, 225
  column-switching technique, 170, 171
  combining separation columns, 170, 171
  comparison of separations with capillary and packed columns, 166, 167
  coupled techniques
    with AES, 196, 197, 228, 229
    with LC, 196, 197
    with MS, 226, 227
  detectors
    electron capture detector (ECD), 172, 173, 226

flame ionization detector (FID), 172, 173

heat conductivity detector (HCD), 170, 171

human nose as, 224

mass-selective detector, 226

thermionic detector (TID), 172, 173, 226

diagram of GC device, 164, 165

error recognition in GC analysis
  baselines, 168, 169
  ghost peaks, 168, 169
  noise, 168, 169
  peak asymmetry, 168, 169
  spikes, 168, 169

gradient trap for trace analysis, 170, 171

headspace technique, 170, 171, 224

inactivation of surfaces, 164

isotherm separation, 168, 169

process analysis using, 214

retention indices, 166, 167

sample injector, 164

sample preparation for, 30

separation column, 164, 165

symbols for, 9, 164, 165

temperature dependence of GC separation, 168, 169

temperature-programmed separation, 168, 169

Gas detector tubes, 36, 37

Gas-permeable membranes, 56, 57, 192, 193

Gas-selective/sensitive electrodes, 56, 57

Gases
  absorption into liquids, 18, 19, 26
  adsorption onto solids, 140, 141
  analysis by NDIR gas analyzer, 215
  sampling of, 18, 19

Gaussian [bell-shaped] curve, 14, 15, 66, 142, 143

Gel electrophoresis, 174, 175, 176, 177, 182, 183

Gel filtration, 140, 141

Gel permeation chromatography, 140, 141, 146, 147, 162, 163
  applications, 162, 163, 220

German silver alloy, analysis of, 202, 203

Glass electrodes [for pH measurement], 54, 55, 56

alkaline errors, 54, 55
Glass membrane, potential diagram for, 54, 55
Glauberite, 78, 79
Glucose, enzymic analysis of, 46, 47
Glutamate, determination of, 46, 47
Gradient trap [GC], 170, 171
Granite, analysis of, 102, 103
Gravimetry, 6, 7, 38–9
    advantages/disadvantages, 38
    concentration range, 10, 11
    procedures, 38, 39
    sensitivity plots, 16, 17
    see also Electrogravimetry; Thermo-gravimetry
GROTRIAN diagram, 92, 93
Gypsum, 78, 79

Half-wave potential [polarography], 62, 63
Halide ions
    potentiometric titration, 58, 59
    semi-quantitative analysis, 36, 37
Haptens, 48, 49
Hard metal, SIMS analysis of, 202, 203
Harmonic vibration, 114, 115
Heat conductivity detector [GC], 170, 171
Heavy metals
    compartmentalization in plants, 220, 221
    determination of, 64, 68, 69, 86, 87, 90, 91, 96, 97, 102, 103, 220, 221
    sample preparation methods used, 24, 28
    see also Cadmium; Copper; Lead; Nickel; Zinc
Helmholtz double layer, 50, 51, 62, 68
Hertz, meaning of term, 104
Hexanone isomers, IR spectra, 116, 117
High-performance capillary electrophoresis (HPCE), 182, 183
High-performance liquid chromatography (HPLC), 8
    applications, 196, 197, 228, 229
    chemical reactor inserted, 160, 161
    comparison with other chromatographic methods, 152, 153, 154
    coupled techniques
        with FTIR, 228, 229

with other chromatographic techniques, 196, 197
detectors
    amperometric detectors, 156, 158, 159
    chemical reaction detector, 160, 161
    flow-through cell for, 156, 158, 159
    noise and drift, 158, 159
    photodiode array detector, 158, 159, 228, 229
    refractive index (RI) detector, 156, 158, 159
    elemental species analysis using, 220, 221
    process analysis using, 214, 215
    reciprocating pump, 156, 157
    relationship of height equivalent to theoretical plate and particle diameter, 158, 159
    sample application valve, 158, 159
    sample preparation for, 30, 31
    structure, 156, 157
    symbol for, 9
    see also Liquid chromatography
High-performance thin-layer chromatography (HPTLC), 148, 149
    see also Thin-layer chromatography
High-pressure incinerator, 22, 23
Hydrocarbon brushes, 162
Hydrogen
    energy-level diagram, 92, 93
    line spectrum, 92, 93
    spin of nucleus, 132, 133
Hydrogen electrodes, 56, 57
Hydrogen fluoride, NMR spectrum, 134, 135
Hydrogen sulfide, precipitation by, 32, 33, 34, 35
8-Hydroxyquinoline, 34, 35

Identification reagents, 32
Illite, 78, 79
Immunoassay, 48, 49
Immunoelectrophoresis, 180, 181
Incineration methods, 22, 23, 24, 25, 26, 27
Indication [of end-point], 40
Indicators
    acid–base titration, 40, 41
    redox titration, 42

Indole, 113
Inductively coupled plasma (ICP)
  in atomic emission spectrometry, 96, 97
  in mass spectrometry, 122, 198, 199
Infrared (IR) region
  correlation table of absorption bands, 120, 121
  fingerprint region, 116, 117
  wavelength range, 104, 105
Infrared (IR) spectrometry, 118, 119
  symbol for, 8, 9
  see also Non-dispersive IR (NDIR) gas analyzer
Infrared (IR) spectroscopy, 2, 3, 6, 114–21
  applications, 120, 121
  cuvettes, 114, 115
  Raman scattering compared with IR absorption, 114, 115, 120, 121
  spectrometer, 114, 115
  structural analysis using, 216, 217
  vibrational modes, 114, 115, 116, 117
Instrumental methods, 6, 7
Instrumental neutron activation analysis (INAA), 200
Interface electrophoresis, 174
Interface reactions, in chemical sensors, 206
Interference level, 14, 15
International Union of Pure and Applied Chemistry (IUPAC), nomenclature for analytical applications, 10
Inverse differential pulse polarography, 68, 69
Inverse voltammetry, 28, 68, 69
Ion exchange
  sample enrichment using, 28, 29
  separation by, 140, 141, 146, 147
Ion-exchange chromatography (IEC), 6, 8, 160, 161
  symbol for, 9
Ion-exclusion chromatography, 146, 147
Ion-pair partitioning, 146, 147
Ion-selective electrodes (ISE), 56, 57, 58, 194
  calibration curves for, 58, 59, 194, 195
Iron
  determination of, 16, 17, 102, 103
  resolution of triplet, 96, 97

wet-chemical detection/separation of, 33, 34, 35, 36, 37, 38
Isoelectric focusing [electrophoresis], 174, 175, 180, 181
Isoenzymes, electrophoresis of, 176, 177
Isotachophoresis, 174, 175, 178, 179

Jablonski diagram, 112, 113

Kaolin, 78, 79
Kieserite, 78, 79
Kjeldahl [decomposition] method, 24
Knapps high-pressure incinerator, 22, 23
Kováts retention index, 166, 167
Kubelka–Munk function, 152

Lambda probe, 206, 208
Lambert–Beer law, 82, 106, 108
Langmuir adsorption isotherm, 146, 147
Lanthanum–arsenazo complex, photometry of, 110, 111
Laser atomic absorption spectrometry (LAAS), 204, 205
Laser spectrometry, 204, 205
Laser-enhanced ionization (LEI) spectrometry, 204, 205
Laser-induced fluorescence (LIF), 204, 205
Lasers, 204
Laue [X-ray diffraction] method, 218, 219
Lead
  compartmentalization in plants, 220, 221
  determination of, 64, 68, 69, 90, 91, 102, 103, 220, 221
  separation of, 33, 34, 35
Limiting concentration, 32, 36
Limiting diffusion current, 63, 65, 67
Lindane [pesticide], GC determination of, 172, 173
Linear non-ideal chromatography, 144, 145
Liquid chromatography (LC), 8, 154–63
  apparatus/devices, 156, 157
  chemically bonded phases, 160, 161
  column-switching technique, 30, 31
  coupled techniques
    with GC, 196, 197
    with TLC, 196, 197

detectors
    amperometric detectors, 158, 159
    chemical reaction detector, 160, 161
    flow-through cell for, 158, 159
    noise and drift, 158, 159
    photodiode array detector, 158, 159
    refractive index detector, 158, 159
    UV/VIS photometric detector, 156
displacement technique, 154, 155
elution technique, 154, 155
gradient elution, 156, 157
isocratic elution, 156, 157
matchbox model, 154, 155
sample preparation for, 30, 31
symbol for, 9
    see also High-performance liquid chro-
        matography
Liquid membrane electrodes, 56, 57, 58
Liquid–gas–liquid enrichment methods,
    28, 29
Liquid–liquid enrichment methods, 28, 29
Liquid–liquid extraction, 6, 140, 141
    with FIA system, 194, 195
    see also Shaking out
Liquids, sampling of, 18, 19
Lithium
    alkaline error of pH glass electrodes, 55
    emission line spectrum, 94, 95
    energy-level/term diagram, 94, 95
Loeweite, 78, 79
Lorentz energy, 124
Low-pressure liquid chromatography
    components of apparatus, 156, 157
    see also Liquid chromatography
L'vov platform [for AAS], 84, 85, 90
Lyman series [for emission spectrum of
    hydrogen], 92, 93

McLafferty rearrangement, 130, 131
Macro-sample, 10, 11
Macroscale accuracy, 38
Magnesium
    determination of, 42, 43, 102, 103
    wet-chemical separation of, 34, 35, 38
Magnetic quantum number, 132
Maleic acid, polarographic determination
    of, 68, 69
Malic acid, enzymic analysis of, 46, 47
Manganese
    determination of, 102, 103

wet-chemical detection/separation of,
    33
Mannose, enzymic analysis of, 46, 47
Mass fragmentography, 226
Mass range, 10, 11
Mass spectrometry (MS), 2, 3, 6, 122–31
    applications, 128–31
    coupled techniques
        with gas chromatography, 226, 227
        with liquid chromatography, 196,
            197
        with resonance ionization spectro-
            metry, 204
    double-focusing mass spectrometer,
        126, 127
    in SIMS, 126, 127, 202, 203
    electron impulse ion source, 122, 123
    focusing methods
        directional focusing, 124, 125
        electrostatic focusing, 124, 125, 126
        magnetic focusing, 124, 125, 126
        quadrupole focusing, 124, 125, 126
    fragmentation diagrams, 122, 123, 128,
        129
    ion separation in, 140, 141
    ionization methods
        chemical ionization (CI), 126, 127
        electron impact (EI) ionization, 122,
            126, 127, 128, 129
        field desorption (FD), 126, 127
        field ionization (FI), 126, 127, 128,
            129
        inductively coupled plasma (ICP–
            MS), 198, 199
    monochromator, 124, 125
    spectrometer make-up, 122, 123
    structural analysis using, 216, 217
    symbol for, 8, 9
    time-of-flight spectrometer, 126, 127
Massmann cuvette [for AAS], 84
Maximum   workplace   concentration
    (LWC), 36
Melt decompositions, 22, 23
Melting points, polymers, 78
Mercury
    determination of, 84
    drop formation in polarography, 62, 63
    separation of, 33, 34, 35
    see also Dropping mercury electrode
Meso-sample, 10, 11

Metal ions, wet-chemical detection/separation of, 32–7
Metal sulfates, decomposition voltages, 52, 53
Methacrylic acid β-dimethylaminomethyl ester, structural analysis of, 216, 217
Methylene group, vibrational modes, 114, 115, 117
Michaelis–Menten equation, 44, 45
Michelson interferometer, 118, 119
Micro-sample, 10, 11
Microorganisms, thermal differentation of, 78, 79
Microwave absorption spectroscopy, electromagnetic spectrum, 105
Microwave decomposition systems, 24, 192
Microwaves, wavelength range, 104, 105
Minimum mass of sample, 20, 21
Mirabilite, 78, 79
Molecular spectroscopic methods, 2, 3, 6, 104–39
   see also Colorimetry; Fluorimetry; Infrared spectroscopy; Mass spectrometry; Nuclear magnetic resonance spectroscopy; Raman spectroscopy; Spectrophotometry
Molecular vibration, 104, 105, 114, 115, 116, 117
Molybdenum, determination of, 102, 103
Monochromatic light, meaning of term, 108
Montmorillonite, 78, 79
Moseley's law, 98
Multi-element method, 10
Multiplicative errors, 16

Nernst distribution, 146
Nernst equation, 42, 50, 56
   expanded equation, 58, 59
Nernst potential, 50, 206
Nernst rod, 114
Neutron activation analysis, 200, 201
Nickel
   determination of, 64, 102, 103
   wet-chemical detection/separation of, 33, 34, 35, 36, 37, 38
Nickel sulfate, decomposition voltage, 53

β-Nicotinamide adenine dinucleotide (NAD), 44, 45
   absorption spectra, 44, 45
β-Nicotinamide adenine dinucleotide phosphate (NADP), 44, 45
Niobium, determination of, 102, 103
Nitrate, determination in water, 194, 195
Nitropropane, NMR spectrum, 136, 137
Non-dispersive IR (NDIR) gas analyzer
   functional principles, 214, 215
   see also Infrared (IR) spectrometry
Non-linear ideal chromatography, 144, 145
Non-linear value-dependent errors, 16
Noradrenaline, determination of, 188, 189
Normal distribution, 14, 15
Nuclear magnetic resonance (NMR) spectroscopy, 2, 3, 6, 132–9
   anisotropy effect, 134, 135
      with aromatic compounds, 134, 135
   ^{13}C-NMR spectroscopy, 138, 139
      with broadband decoupling, 138, 139
      with off-resonance decoupling, 138, 139
      stretching/widening of spectrum, 138, 139
      without proton decoupling, 138, 139
   chemical shift, 132, 133
   inductive effect, 132, 133
   principles, 132, 133
   proton off-resonance decoupling, 138, 139
   spectrometer, 136, 137
   spin–spin coupling, 134, 135
   structural analysis using, 216, 217
   symbol for, 8, 9
Nuclear resonance, 132, 133

Oils
   analysis of, 102, 103
   determination in soils/water, 116
Optical methods, 6, 7
   see also Atomic...; Molecular spectroscopic methods
Optical sensors, 208, 209
Overvoltage [in electrolysis], 50, 52
Oxalates, thermal analysis of, 74, 75, 78, 79
Oxidation–reduction titrations, 42
   see also Redox titrations

Oxide sensors, 206, 207
Oxidimetric titrations, 42
Oxygen bomb, 26, 27
Oxygen sensors, 208, 209
Oxygen stream, combustion in, 26, 27

Paper electrophoresis, 174, 175, 176, 177
Paramagnetic oxygen measurement, 210, 211
Parr bomb, 22, 23
Particle charge, separation using, 140, 141
Particle mass/shape/size, separation using, 140, 141
Particle size, and minimum sample mass, 20, 21
Partition chromatography, principles, 154, 155
Partitioning
  separation using, 140, 141
  see also Chromatography; Extraction; Liquid–liquid extraction; Shaking out; Solid–liquid extraction
Paschen series [for emission spectrum of hydrogen], 92, 93
pD value, 32
Penetration reaction [electrochemistry], 50, 62
Pentanone isomers, comparison of MS spectra, 130, 131
Peroxidase, separation of, 162, 163
Pesticide analysis, 30, 48, 172, 226–9
pH measurement
  alkaline errors of glass electrodes, 54, 55
  optosensor, 208, 209
  in process analysis, 210, 211
  pH-measuring techniques, in potentiometry, 54, 55
Phase conversion, sample enrichment using, 28, 29
Phenolphthalein [indicator]
  mode of action, 40, 41
  transition range [pH], 41
Phosphates
  determination of, 16, 17, 184, 185, 192, 193
  thermal analysis of, 78, 79
Phosphorescence, 112

Phosphorus, determination of, 90, 91, 96, 97
Phosphorus–nitrogen detector (PND), 172
  see also Thermionic detector
Photodiode array detectors, 158, 159, 228, 229
Photoelectric effect, 100
Photoelectron spectrometry (PES)
  principles, 204, 205
  spectra for copper surfaces, 204, 205
  X-ray photoelectron spectrometry, 204, 205
Photometers
  filter photometers, 108, 109
  process, 210, 211
  spectrophotometry, 108, 109
Photometry
  automated system, 186
  calibration curves, 16, 17, 110, 111
  concentration range, 10, 11
  detection/determination limits, 14, 15
  immunoassays, 48, 49
  NAD–NADH/NADP–NADPH, 44, 45
  sample preparation for, 24
  two-component analysis, 110, 111
  see also Flame atomic emission spectrometry; UV–VIS spectrophotometry
Photomultipler, 120
Planck's relationship, 104
Plant materials
  compartmentalization of heavy metals in, 220, 221
  residue analysis of, 226, 227
Plasma excitation, in emission spectrometry, 96–7
Polarography, 6, 62, 63, 64–9
  differential pulse polarography, 62, 66, 67
  pulse polarography, 66, 67
  sample preparation for, 24
  see also Differential pulse...; Inverse differential pulse...; Pulse...; Tast a.c. polarography
Polychromatic light, meaning of term, 108
Polycyclic aromatic hydrocarbons (PAHs), separation from other compounds, 30

Polyethylene, thermal analysis of, 74, 75, 78, 79
Poly(ethylene terephthalate), thermal analysis of, 80, 81
Polymers, thermal analysis of, 74, 75, 78, 79, 80, 81
Polytetrafluoroethylene, thermal analysis of, 74, 75, 78, 79
Potassium, alkaline error of pH glass electrodes, 55
Potentiometric titration, 58, 59
Potentiometry, 6, 54–9
    applications, 58–9
    concentration range, 10, 11
    electrodes used, 56–7
    pH-measuring techniques, 54–5
Potentiostatic coulometry, 70, 71
ppm/ppb/ppt/ppq, meaning of terms, 10, 11
Precession, 132
Precipitating reagents, 34, 35, 38
Precipitation, 38, 140, 141
    symbol for, 9
    see also Co-precipitation
Precipitation analysis, 7
    sources of error, 38
    see also Gravimetry
Precipitation titrations, 40, 42, 43
Precision of procedure, 14
Preconcentration methods, 28–9
Pressure (melt) decompositions, 22, 23
Problem definition, 4, 5
Procedural protocols, 6
Procedural steps, 8
    in process analysis, 212, 213
    symbols for, 9
Process analysis, 2, 3, 210–15
    oxygen measurement, 210, 211
    pH measurement, 210, 211
    by photometry, 210, 211
    procedural steps in, 212, 213
    by refractometry, 210, 211
    signal forms, 210, 211
    by X-ray fluorescence spectrometry, 214, 215
Propane, NMR spectrum, 136, 137
Protein complexes, analysis of, 198, 199
Protein mixtures, separation of, 176, 177, 180, 181, 182
Pulse polarography, 66, 67

Pumping of gases
    separation by, 8
    symbol for, 9

Quadrupole mass spectrometry, 124, 125, 126
    coupled with LC, 196, 197
    with inductively coupled plasma (ICP–MS), 198, 199
    structure, 198, 199
    see also Mass spectrometry
Qualitative analysis, 2, 3
    working methods, 32, 33
Quantitative analysis, 2, 3
Quinhydrone electrode, 56
Quinine, quantitative TLC analysis of, 152, 153

Radio waves, wavelength range, 104, 105
Radiochemical methods, 200–1
Radioimmunoassays, 200
Radio-thin-layer chromatography, 200, 201
Raman scattering, compared with infrared absorption, 114, 115, 120, 121
Raman spectroscopy, 120, 121
Random errors
    sources, 12, 13
    statistical errors and, 14, 15
Rayleigh scattering, 114, 115
Redox electrodes, 56, 57
Redox titrations, 40, 42, 43
Reducing sugars, separation of, 182, 183
References listed, 230–1
Refractive index (RI) detectors, 158, 159
Refractometer, process, 210, 211
Refuse, sample preparation for, 20, 21, 90
Refuse drainage water, lead in, 90, 91
Relaxation, nuclear resonance, 132, 133
Remission measurement/analysis, 152, 208
Representativeness of sample, 20
Resonance energy, 132
Resonance ionization mass spectrometry (RIMS), 204, 205
Resonance ionization spectrometry (RIS), 204, 205
Retro-Diels–Alder (RDA) cleavage, 130, 131

Reversed-phase chromatography, 162, 163, 196
  separation mechanism, 146, 147
Reversed-phase thin-layer chromatography, comparison with HPLC, 152, 153
Riboflavin [vitamin B$_2$], calibration curve for, 16, 17
Ribose, MS spectra, 128, 129
Rocket technique [immunoelectrophoresis], 180, 181
Rocks, analysis of, 102, 103
Rohrschneider constants, 166
Rowland circle [in emission spectrometer], 96, 97
Rubidium, determination of, 102, 103

Sample cleanup procedures, 30–1, 194, 195, 226, 227
Sample decomposition methods, 22–7, 222
Sample enrichment methods, 28–9
Sample preparation, 4, 5, 18–31, 102, 103
Sample reduction, 20
Sample size, selection of method affected by, 10, 11
Sample splitting, 20, 21
Sample stabilization, 18
Sampling methods, 18–21
Schellbach strips [on burettes], 40, 41
Schöniger flaks, combustion in, 26, 27
Schottky diode, 206, 207
Scintillation counters, 100, 101
Seawater, enrichment of trace elements from, 28
Secondary ion mass spectrometry (SIMS) principles, 126, 127, 202, 203
  signal generation in, 202, 203
Sedimentation analysis, 140, 141
Selective reagents, 32, 33, 34, 35, 36, 37
Selenium, determination of, 84, 90, 91, 102, 103
Semiconductor [chemisorption] sensors, 206
Semiconductor detectors [X-ray fluorescence], 100, 101
Semimicro-sample, 10, 11
Sensitivity of process, 16, 17
Sensors
  meaning of term, 206

see also Electrodes
Sensory analysis, 224
Separation methods, 6, 7, 8, 140–83
  by particle charge differences, 140, 141
  by particle effects, 140, 141
  by partitioning, 140, 141
  substance-transformation methods, 140, 141
  symbols for, 9
  by vapor pressure differences, 140, 141
Separation procedures [wet chemistry], 34, 35
Serum proteins, electrophoretic separation of, 176, 177, 178, 179
Sewage sludge, dried, sample preparation of, 26
Sewage water
  analysis of, 90, 91, 194, 195
  BOD determination, 212, 213
Shaking out
  symbol for, 9
  see also Liquid–liquid extraction
Sieving
  separation by, 8
  symbol for, 9
Silicon
  determination of, 102, 103
  emission spectrum, 86, 87
Silver
  titrimetric determination of, 42, 43
  wet-chemical detection/separation of, 37, 38, 39
Silver sulfate, decomposition voltage, 53
Sniffing test [GC], 224
Sodium, alkaline error of pH glass electrodes, 55
Sodium dodecylsulfate (SDS) electrophoresis, 180
Sodium salts, dissociation continua, 88, 89
Sodium-selective electrode, calibration curve for, 58, 59
Soils
  determination of oils in, 116
  heavy metals in, 220, 221
  sampling of, 20, 21
Solid–liquid extraction, 6, 140
Solid-body analysis, 202, 204, 205
Solid-phase extraction, 30, 31
Solids, sampling of, 20, 21

Soya proteins, separation of, 176, 177
Soybean oil, analysis of, 90, 91
Special methods, 200–29
Specific conductivity, 60
  see also Conductometry
Spectrophotometry, 2, 3, 6, 108–11
  applications, 108, 109, 110, 111
  distribution of relative measuring errors, 110, 111
  sources of error, 108, 109
  structure of photometers, 108, 109
  wavelength-dependent absorption, 108, 109
Spectroscopic methods
  see Infrared spectroscopy; Mass spectrometry; Nuclear magnetic resonance spectroscopy; Raman spectroscopy
Spectroscopic methods
  see UV–VIS spectrophotometry
Spin quantum number, 132
Spot plate analysis, 36, 37
Standard deviation, 14, 20
Statistical evaluation of results, 14–17
Steel alloys, analysis of, 102, 103
Stokes' law, 112, 174
Stokes [spectral] region, 114, 115
Stopped-flow analysis, 190, 191
Strategic approach to analysis, 4, 5
Streptomyces spp., 78, 79
Strontium
  determination of, 102, 103
  separation of, 33, 34, 35
Structural analysis, 2, 3, 216–19
  qualitative/semiquantitative, 216, 217
Structural models, 218, 219
Styrene oligomers, separation of, 162, 163
Sublimation, separation by, 140, 141
Submicro-sample, 10, 11
Sugars
  electrophoretic separation of, 182, 183
  MS spectra, 128, 129
Sulfates
  decomposition voltages in electrolysis, 53
  thermal analysis of, 78, 79
Sulfur, determination of, 102, 103
Surface analysis, 2, 3, 202–5
Sweep co-distillation (SCD), 30, 31, 226

Symbols
  analytical methods, 9
  gas chromatography equipment, 164, 165
  procedural steps, 9
  separation methods, 9
Systematic errors
  sources, 12, 13, 16, 22, 102
  statistical errors and, 14, 15

Talc, 78, 79
Tast a.c. polarography, 68, 69
Tea extracts, analysis of, 200, 201
Tellurium, determination of, 84
Term diagrams, 105
  fluorescence, 112, 113
  for hydrogen, 92, 93
  laser-induced radiation, 205
  for lithium, 94, 95
  nuclear resonance, 135, 137
  X-ray fluorescence, 98, 99
Test papers/strips
  with color comparison, 36, 37
  for halide ions, 36, 37
  for metal ions, 36, 37
Tetrabromomethane, GC determination of, 172, 173
n-Tetradecane, MS spectra, 128, 129
Thallium, polarographic analysis of, 64
Thermal analysis, meaning of term, 72
Thermal analytical methods, 72–81
Thermal conductivity coefficients, 171
Thermionic detector (TID), 172, 173, 226
Thermogravimetric analysis (TGA)
  classification of TGA curves, 74, 75
  polymer characterization, 74, 75
  simultaneous determination of coprecipitates, 74, 75
  water release from oxalates, 74, 75
Thermogravimetry (TG), 72, 73, 74–5
Thermomechanical analysis (TMA), 72, 73
Thermometric titration, 72, 73
Thermomicro-separation, transfer and application (TSA) procedure, 30, 31
Thiamine [vitamin B_1], determination of, 160, 161
Thin-layer chromatography (TLC), 8, 148–53
  automatic sample application, 150, 151

calibration curves, 152, 153
comparison with HPLC, 152, 153
coupled with liquid chromatography, 196, 197
development chambers, 148, 149
development techniques, 150, 151
elution activity of different solvents, 148, 149
elution strength in solvent mixture, 148, 149
fluorescence indicators used, 152, 153
immersion chamber for reagent application, 150, 151
kinetic theory, 142, 143
multiple TLC, 150
quantitative TLC, 152, 153
radiochemical method, 200, 201
retention factor value, 150, 151
    spot penetration depth influenced by, 150, 151
sample preparation for, 30, 31
separation procedure, 148, 149
spot penetration depth, effect of retention factor value, 150, 151
symbol for, 9
two-dimensional TLC, 150
Thorium, determination of, 102, 103
Time-of-flight mass spectrometer, 126, 127
Tin
    determination of, 84, 102, 103
    separation of, 33, 34
Tin dioxide chemisorption sensor, 206, 207
Tiselius [electrophoresis] method, 174
Titanium, determination of, 102, 103
Titrimetry, 6, 7, 40–3
    acid–base, 40, 41
    argentimetric precipitation, 42, 43
    complexometric, 40, 42, 43
    concentration range covered by, 10, 11
    coulometric, 70, 71
    flow injection analysis and, 190, 191
    potentiometric, 58, 59
    precipitation, 40, 42, 43
    redox, 40, 42, 43
    thermometric, 72, 73
Tölg bomb, 22, 23
Total inorganic carbon (TIC) content, determination of, 222

Total organic carbon (TOC) content, 222, 223
    determination of, 222, 223
Trace element analysis, 68, 69, 90, 200
Transmission spectra, 108, 109
Triazine pesticides, determination of, 48, 49, 226, 227, 228, 229
Trichloromethane, IR transmission spectrum, 116, 117
2,2,4-Trimethylpentane, NMR spectrum, 138, 139
Two-dimensional chromatography, 196

Ultracentrifugation, 140, 220, 221
Ultrafiltration, 220
Ultramicro-sample, 10, 11
Ultraviolet (UV) region, wavelength range, 104, 105
Urates, thermal analysis of, 78, 79
Urease electrode, 56
Urinary stones, thermal analysis of, 78, 79
Urine analysis, 86, 87
Urotropine (hexamethylenetetramine), 32, 33, 34, 35
UV decomposition device, 24, 25, 222
UV–VIS spectrophotometry, 2, 3, 108–11
    applications, 108, 109, 110, 111
    calibration for, 110, 111
    effect of spectral bandwidth, 110, 111
    photometers used, 108, 109
        in chromatography detectors, 156
    sources of error, 108, 109
    structural analysis using, 216, 217
    two-component analysis, 110, 111
    wavength range, 105

Van Deemter equation, 144, 145, 158
Van der Waals forces, 134, 135
Vanadium, determination of, 102, 103
Vapor pressure differences, separation using, 8, 140, 141
Vibrational modes, IR spectra, 114, 115, 116, 117
Visible-light spectrometry, 108
    wavength range, 106, 107, 108
    see also UV–VIS spectrophotometry
Vitamins, determination of, 16, 17, 160, 161
Volatilization, separation by, 140, 141

Voltage–current curves
  electrolysis cells, 50, 51
  galvanic cells, 50, 51
  see also Current–voltage curves
Voltammetry, 62–3
  concentration range, 10, 11
  sample preparation for, 24
Volumetric analysis, 6, 7
  see also Titrimetry

Waste water, analysis of, 96, 97
Water
  analysis of, 68, 69, 90, 91, 120, 121,
    194, 195, 222–3
  determination of oils in, 116
  hardness determination of, 42
  IR absorption band, 120, 121
  organic halo-compounds in, 222, 223
  sampling from, 18, 19
  total organic/inorganic content, 222,
    223
Wavelength-dispersive X-ray fluores-
    cence analysis (WDXRFA), 100,
    101
  applications, 102, 103
Wet incineration automat, 24, 25
Wet-chemical methods, 6, 7, 32
Wheatstone bridge circuit, 170, 171
Wickhold apparatus, combustion in, 26,
    27
Witt's suction pot, 39
Working methods, 32, 33
Workstation functions [in automated ana-
    lysis], 184, 185

X-ray diffraction methods, 218, 219
  Debye–Scherrer procedure, 218, 219
  goniometer method, 218, 219
  Laue method, 218, 219
  revolving crystal method, 218, 219
X-ray emission spectrometry, 92
X-ray fluorescence analysis (XRFA), 6,
    98–103
  applications, 102, 103, 214, 215
  detectors used, 100, 101
  energy-dispersive XRFA, 100, 101,
    102, 103
  primary processes, 98, 99
  principle behind excitation, 98, 99
  sample preparation for, 29, 102, 103
  secondary processes, 98, 99
  wavelength-dispersive XRFA, 100,
    101, 102, 103
X-ray photoelectron spectrometry, 204,
    205
X-ray spectrometers, 100, 101
X-ray tubes, 100, 101

Zeeman effect, 88, 89
Zinc
  determination of, 68, 69, 102, 103
  mobility from soils, 220, 221
  wet-chemical detection/separation of,
    33
Zinc sulfate, decomposition voltage, 53
Zirconium, determination of, 102, 103
Zone electrophoresis, 174, 175
  with immunodiffusion, 180, 181
  principles, 174, 175